ANALYSIS OF MULTICONDUCTOR TRANSMISSION LINES • *Clayton R. Paul*

INTRODUCTION TO ELECTROMAGNETIC COMPATIBILITY • *Clayton R. Paul*

INTRODUCTION TO HIGH-SPEED ELECTRONICS AND OPTOELECTRONICS • *Leonard M. Riaziat*

NEW FRONTIERS IN MEDICAL DEVICE TECHNOLOGY • *Arye Rosen and Harel Rosen (eds.)*

NONLINEAR OPTICS • *E. G. Sauter*

FREQUENCY SELECTIVE SURFACE AND GRID ARRAY • *T. K. Wu (ed.)*

ACTIVE AND QUASI-OPTICAL ARRAYS FOR SOLID-STATE POWER COMBINING • *Robert A. York and Zoya B. Popović (eds.)*

OPTICAL SIGNAL PROCESSING, COMPUTING AND NEURAL NETWORKS • *Francis T. S. Yu and Suganda Jutamulia*

Superconductor Technology

Superconductor Technology

Applications to Microwave, Electro-Optics, Electrical Machines, and Propulsion Systems

A. R. JHA

A WILEY-INTERSCIENCE PUBLICATION
JOHN WILEY & SONS, INC.
NEW YORK / CHICHESTER / WEINHEIM / BRISBANE / SINGAPORE / TORONTO

This book is printed on acid-free paper.

Copyright © 1998 by John Wiley & Sons, Inc. All rights reserved.

Published simultaneously in Canada.

No part of this publication may be reproduced, stored in a retrieval system or transmitted in any form or by any means, electronic, mechanical, photocopying, recording, scanning or otherwise, except as permitted under Sections 107 or 108 of the 1976 United Staes Copyright Act, without either the prior written permission of the Publisher, or authorization through payment of the appropriate per-copy fee to the Copyright Clearance Center, 222 Rosewood Drive, Danvers, MA 01923, (508) 750-8400, fax (508) 750-4744. Requests to Publisher for permission should be addressed to the Permissions Department, John Wiley & Sons, Inc., 605 Third Avenue, New York, NY 10158-0012, (212) 850-6011, fax (212) 850-6008, E-Mail: PERMREQ @ WILEY.COM.

Library of Congress Cataloging-in-Publication Data:

Jha, A. R.
    Superconductor technology : applications to microwave, electro
–optics, electrical machines, and propulsion systems / A.R. Jha.
       p. cm. — (Wiley series in microwave and optical engineering)
    Includes bibliographical references and index.
    ISBN 0-471-17775-X (cloth : alk. paper)
    1. Superconductors. 2. Microwave devices—Materials.
3. Electrooptical devices—Materials. 4. Electric machinery-
–Materials. 5. Electric propulsion—Equipment and supplies-
–Materials. I. Title. II. Series.
TK454.4.S93J43 1998
621.3′5—dc21   97–44439

Printed in the United States of America.

10 9 8 7 6 5 4 3 2 1

Contents

Preface   xv

1 Phenomenology and Theory of Superconductivity   1

    1.1 History of Superconducting Materials   2
        1.1.1 History of Low-Temperature Superconductor (LTSC) Materials   2
        1.1.2 History of High-Temperature Superconductor (HTSC) Materials   3
    1.2 Unique Features of Superconductors   4
    1.3 Microscopic Structures of HTSC Compounds   5
    1.4 Phase Formations of Crystalline Structures   7
    1.5 Conduction   7
    1.6 Anisotropic Characteristics of Superconductors   9
    1.7 Superconducting Coherence   9
    1.8 Superconducting Energy Gap   9
    1.9 Grain Boundaries   10
    1.10 Pinning Energy   14
    1.11 Critical Temperature   14
    1.12 Lattice Vibrations   15
    1.13 Impact of Magic Angle on Transition Temperature   15
    1.14 Integration of Superconducting Technology With Other Technologies   17
    Summary   18
    References   18

2 Superconductor Forms and Their Critical Microwave Properties   20

    2.1 General Properties and Applications   20
    2.2 Superconducting Powder   21

	2.3 Superconducting Bulk Materials	21
	2.3.1 Optimization of Processing Conditions for Bulk Ceramics	22
	2.4 Superconducting Wires	23
	2.5 Superconducting Tapes	24
	2.6 Superconducting Films	26
	2.6.1 Superconducting Thin Films	26
	2.6.2 Superconducting Thick Films	30
	2.7 Critical Microwave Properties	33
	2.7.1 Critical Current Density	34
	2.7.2 Radio-Frequency Surface Resistance	34
	2.7.3 Penetration Depth	36
	2.7.4 Meissner Effect	37
	Summary	38
	References	38
3	**Superconducting Substrate Materials**	**40**
	3.1 Classification of HTSC Dielectric Substrates	40
	3.2 Soft Dielectric Substrates	41
	3.2.1 Electrical Properties	41
	3.2.2 Relative Dielectric Constants and Loss Tangents	41
	3.3 Hard Dielectric Substrates	45
	3.3.1 Critical Electrical Properties	46
	3.3.2 HTSC Substrate Requirements at MM-Wave Frequencies	50
	3.3.3 Projection of Insertion Losses	54
	3.4 The Latest Hard Substrate Materials for Microwave Circuit Applications	55
	3.5 Metallic Substrates for High-Power Circuits	57
	Summary	58
	References	58
4	**Application of Superconducting Technology to Passive Components**	**60**
	4.1 Derivation of Surface Impedance	60
	4.2 Propagation Characteristics of Superconducting Microstrip Transmission Lines	64
	4.3 Attenuation Constant (Ac) for a Superconducting Line	66
	4.4 dc Critical Current Density in the Microstrip Line	67
	4.5 Passive Microwave Components Using HTSC Technology	68
	4.5.1 Frequency Multiplexing Filters	69
	4.5.2 Bandpass Filters	70

	4.5.3	HTSC Low-Pass Filter Using a Coplanar Waveguide (CPW)	72
	4.5.4	Superconducting MM-Wave Filters	73
4.6	Superconducting Delay Lines	76
4.7	Superconducting Resonators	78
4.8	Superconducting Shields	79
4.9	Signal-Processing Devices	79
4.10	Application of HTSC Technology to IR Detectors	80
	4.10.1	Mercury Cadmium Telluride (Hg:Cd:Te) Detectors	80
	4.10.2	Indium Antimonide (InSb) Detectors	81
	4.10.3	Indium Gallium Arsenide (InGaAs) Detectors	81
	4.10.4	Gallium Arsenide (GaAs) and Aluminum Gallium Arsenide (AlGaAs) Detectors	81
4.11	Superconducting Radio-Frequency Antennas	82
	4.11.1	High-T_c Superconducting Electrically Short Dipole Antennas	82
	4.11.2	Superconducting Loop and Helical Printed-Circuit Antennas	83
	4.11.3	MM-Wave Superconducting Anntennas	85
Summary		86
References		86

5 Applications of Superconducting Thin Films to Active Rf Components and Circuits — **88**

5.1	Phase Shifters Using Superconducting Technology	89
5.2	Distributed Josephson Inductance (DJI) Phase Shifter	89
	5.2.1 Design Aspects of the HTSC DJI Phase Shifter	90
	5.2.2 SQUID Geometry and Dimensional Parameters	90
	5.2.3 Operating Principle	92
	5.2.4 Conditions for Optimum Performance	93
	5.2.5 Performance Limitations	93
	5.2.6 Performance Capabilities of Phased Arrays Using Superconducting Phase Shifters	94
5.3	HTSC/Ferroelectric Phase Shifters	95
	5.3.1 Critical Design Aspects	95
	5.3.2 Performance Capabilities and Limitations	95
5.4	Precision Analog Superconducting Microwave (PASM) Phase Shifter Using YBCO Thin Films	96
	5.4.1 Design Aspects and Performance Capabilities	97
	5.4.2 Mathematical Expressions to Compute Phase Shift	98

5.5	Hybrid Design of a Nonreciprocal Phase Shifter Using YIG-Toroid and the Superconducting Transmission Line	98
	5.5.1 Design Aspects and Performance Capabilities	98
	5.5.2 Ferrite Requirements for a Superconducting, Hybrid Phase Shifter	100
5.6	Superconducting Magnetically Tunable Filters Using Ferrites	102
	5.6.1 Superconducting YIG-Tuned Bandpass Filter	102
	5.6.2 Magnetically Tunable E-Plane Filter Using Superconducting Technology	103
	5.6.3 Superconducting Ferrite Circulators and Isolators	104
	5.6.4 Ferrite Material Requirements for Superconducting Circulators and Isolators	105
	5.6.5 Impact of Cryogenic Operation on Ferrite Losses	107
	5.6.6 Impact of Cryogenic Temperature on Anisotropic Field and Linewidth of YIG Materials	108
	5.6.7 Impact of Cryogenic Temperature on Resonance Frequency	110
	5.6.8 Effective Gyromagnetic Radio (g_{eff}) at Cryogenic Operations	110
5.7	Cryogenic Windows for High-Power MM-Wave Sources	110
5.8	Superconducting Microwave Switch for Control Applications	114
	5.8.1 Design Aspects and Structural Details	114
	5.8.2 Performance Capabilities	114
Summary		115
References		115

6 Performance Improvement of Solid-State Devices at Cryogenic Temperatures — 116

6.1	Microwave Diodes	116
6.2	Varactor Diodes	117
	6.2.1 Improvement in Diode Insertion Loss	118
	6.2.2 Varactor Leakage	119
	6.2.3 Varactor Noise Figure	119
	6.2.4 Thermal Performance and Power Dissipation	119
	6.2.5 Improvement in Diode Cutoff Frequency and Quality Factor	120
6.3	PIN Diodes	121
	6.3.1 Power Dissipation	121
6.4	GUNN Diodes	122
6.5	Improved Reliability of Diodes at Cryogenic Temperatures	122

6.6	Improvement of Thermal and Mechanical Properties of Potential Microwave Substrate Materials	123
	6.6.1 Improvement in Thermal Characteristics	123
	6.6.2 Structural Integrity of Cryogenically Cooled Dielectric Substrates	124
6.7	Heterojunction Bipolar Transistors (HBTs)	125
	6.7.1 Performance Improvement of Cryogenically Cooled GaAs HBTs	125
	6.7.2 Cryogenically Cooled HBTs for Optoelectronic Applications	125
	6.7.3 Performance of a Cryogenically Cooled Pseudo-HBT	126
	6.7.4 Cryogenically Cooled Dielectric-Base HBT (DB-HBT)	126
6.8	Transistors for Microwave/MM-Wave Applications	128
	6.8.1 MESFET Devices	128
	6.8.2 Modulation-Doped FETs (MODFETs) and High Electron Mobility Transistors (HEMTs)	133
6.9	HEMT and p-HEMT Devices	136
	6.9.1 Unique Capabilities	136
	Summary	139
	References	141

7 Application of Superconductor Technology to Components Used in Radar, Communication, Space, and Electronic Warfare — 142

7.1	Cryogenically Cooled Solid-State Amplifiers	142
	7.1.1 Performance of Superconducting GaAs Low-Noise Amplifiers	143
	7.1.2 HTSC Heterojunction Bipolar Transistor (HBT) Power Amplifiers	143
	7.1.3 Noise Performance of GaAs HBT Amplifiers at Cryogenic Temperatures	147
	7.1.4 Cryogenically Cooled HEMT, p-HEMT, and DH-HEMT Amplifiers	148
	7.1.5 Cryogenically Cooled Operational Amplifiers	149
	7.1.6 Cryogenically Cooled Parametric Amplifiers (PARAMP)	150
7.2	Antennas	153
	7.2.1 Cryogenically Cooled, Electrically Small, Printed-Circuit Antennas Using Meander-Line Radiating Elements	153
	7.2.2 Cryogenically Cooled, Thick-Film, Printed-Circuit Antenna	155
7.3	Microwave Filters	159
	7.3.1 Low-Power Microwave Filters	159
	7.3.2 High-Power Microwave Filters	159

7.3.3 Advantages of Using HTSC Technology	161
7.4 Delay Lines	161
7.4.1 Wideband-Printed Circuit (WPC) Delay Lines	161
7.5 Cryogenically Cooled Frequency Multipliers	163
7.5.1 Conversion Efficiency	163
7.6 Cryogenically Cooled Mixers	164
7.7 Cryogenically Cooled Rf Sources	166
7.7.1 Rf Oscillators Using Superconducting Ring Resonators	166
7.7.2 Performance Capability Using Solid-State Devices	168
7.7.3 Capability of Superconducting Hybrid Rf Oscillator	169
7.8 MM-Wave Traveling Wave Tube Amplifier (TWTA) Design	170
7.9 Cryogenically Cooled Maser System	170
7.10 Superconductive Free Electron Maser System	172
7.11 Cryogenically Cooled Microwave Receivers	172
7.11.1 Cryogenically Cooled Channelized Receivers	173
7.11.2 Superconducting Electronic System Measurement (ESM) Receiver	174
7.11.3 Cryogenically Cooled Wideband Compressive Receivers	174
7.12 Cryogenically Cooled MM-Wave Surveillance Receivers	177
7.13 Superconducting Analog-to-Digital Converter (ADC)	177
7.13.1 Critical Elements of an ADC	178
7.13.2 Performance of a High-Resolution, High-Sensitivity ADC Using LTSC Technology	178
7.13.3 Superconducting Sigma–Delta ADC Architecture Using Single Flux Quantum (SFQ) Technique	181
7.13.4 Superconducting Digital Radio-Frequency Memory (DRFM)	182
7.14 Superconducting Packet Switches for Communication Systems	183
7.15 Cellular-Base Stations	184
7.16 Space Communication Systems	184
7.17 Radar Systems	185
7.17.1 Cryogenically Cooled FM-CW Radar Capability	185
7.17.2 Phased-Array Radar Systems	186
Summary	188
References	188

8 Applications of Superconducting Technology to Electrooptical Components and Systems   **189**

    8.1 Detectors: Photon, Quantum, and Optical   189

    8.1.1 Performance of Cryogenically Cooled Photon Detectors 190
    8.1.2 Cryogenically Cooled IR Detectors 191
    8.1.3 Quasi-Optical SIS Detectors 193
  8.2 Superconducting Bolometers 196
    8.2.1 Superconducting Hot Electron Bolometer 198
    8.2.2 Superconducting Hot Electron Microbolometer 198
  8.3 Semiconductor Lasers 199
    8.3.1 Cryogenically Cooled InGaAsP/InP Semiconductor Lasers 200
    8.3.2 Cryogenically Cooled InAsSb/AlAsSb and InAsSb/InAsSbP Lasers 201
    8.3.3 Cryogenically Cooled Quantum-Cascade Lasers (QCL) 202
    8.3.4 Cryogenically Cooled Diode Laser With Low Threshold Current 202
  8.4 Solid-State Diode Pump Laser 202
  8.5 IR Imaging Camera 204
  8.6 Staring IR Camera 205
  8.7 Monolithic Charge Transfer Devices 205
  8.8 IR Line Scanner 206
  8.9 Forward-Looking Infrared (FLIR) System 206
  8.10 IR Sensor Using Cryogenically Cooled Multichannel CCDs 207
  8.11 SIS Quasi-Particle Receiver Mixer 207
  8.12 Quasi-Optical SIS (QOSIS) Receiver 209
  8.13 Optical Modulator 210
  8.14 IR Communication Transmitter Using Cryogenically Cooled Light-Emitting Diodes 210
  8.15 Superconducting Scanning SQUID Microscope 212
  Summary 212
  References 212

9 **Applications of LTSC and HTSC Technology to Medical Diagnostic Equipment** **214**
  9.1 Performance Capabilities and Design Aspects of SQUIDs 215
  9.2 Derivation of Critical Performance Parameters of a SQUID 216
    9.2.1 Sensitivity 216
    9.2.2 Flux Density Noise Level ($\sqrt{S_\phi}$) 216
    9.2.3 Flux Noise Level (S_ϕ) 216
    9.2.4 Bandwidth and Slew Rate 217
    9.2.5 Power Spectral Density and Energy Efficiency 217
  9.3 Capabilities of Various Superconducting SQUID Magnetometers 219
    9.3.1 RF-SQUIDs 220
    9.3.2 DC-SQUIDs 225

9.3.3 Multichannel DC-SQUID Magnetometers	231
9.3.4 DC-SQUID Magnetometers for Magnetocardiography	233
9.4 SQUID Gradiometer Systems Using Low-Temperature Superconducting Technology	233
9.4.1 First-Order Gradiometers	234
9.4.2 Magnetically Shielded Second-Order DC-SQUID Gradiometers	235
9.4.3 RF-SQUID-Based Second-Derivative Gradiometers for Neuromagnetic Field Measurements	235
9.4.4 Portable, Low-Temperature SQUID Gradiometers	235
9.5 SQUID Magnetometers and Gradiometers Using High-Temperature Superconductor Technology	238
9.5.1 DC-SQUIDs	238
Summary	239
References	239

10 Application of Superconducting Technology to Generators, Motors, and Transmission Lines   241

10.1 Development History of Superconducting Machines	242
10.2 Advantages of Superconducting Machines	242
10.2.1 Current Capability and Rating	244
10.2.2 Improvement in Efficiency at Cryogenic Temperatures	245
10.3 Superconducting Synchronous Generator	245
10.3.1 Thermal Radiation and Damper Shields	247
10.3.2 Stator Windings	249
10.4 Design Aspects for Efficient Superconducting Machines	249
10.5 Critical Elements of Large Superconducting AC Machines	250
10.5.1 Potential Advantages	251
10.5.2 Critical Design Aspects	251
10.5.3 Cooling Requirements	252
10.5.4 Performance and Cost Comparison for Large Conventional and Superconducting Generators	254
10.5.5 History of Superconducting Turbogenerator Projects	255
10.6 Superconducting Generators for Airborne Applications	255
10.7 Superconducting Motors	257
10.8 Applications of Superconducting Synchronous Machines for Ship-Propulsion Systems	258
10.9 Superconducting Electrodynamic Levitation Systems	262
10.9.1 Operating Principle of a Levitated Train	262
10.9.2 Superconducting Propulsion System	262
10.9.3 Programs Undertaken on Levitation Systems	264
10.10 Superconducting Transmission Lines and Cables	266

		10.10.1 Design Aspects	267
		10.10.2 Cooling Systems	267
	10.11	Superconducting Components for Electrical High-Power Systems	269
		10.11.1 Superconducting Fault Current Limiter	269
		10.11.2 Superconducting Current Transfer Device (SCTD)	271
		10.11.3 Superconducting Bearing	273
	Summary		273
	References		274
11	**Cryogenic Refrigerator Systems**		**275**
	11.1	Critical Requirements and Operational Characteristics	276
	11.2	Capabilities of Commercially Available Refrigerators	276
		11.2.1 Dilution-Magnetic (DM) Cryocoolers	277
		11.2.2 Collins–Helium Liquefier (CHL)	277
		11.2.3 Gifford–McMahan (GM) Refrigerators	277
		11.2.4 GM/JT Refrigerators	278
		11.2.5 Stirling Cryocoolers	280
		11.2.6 Self-Regulated Joule–Thompson Cryocooler	281
	11.3	Closed-Cycle Cryogenic (CCC) Refrigerator	281
	11.4	Cooling Schemes Used by Various Cryocoolers	284
	11.5	High-Temperature Cooling System for Sonar Transmitters	285
		11.5.1 Cooling Power Levels at High Superconducting Temperatures	287
	11.6	Progress in Miniaturized Cryocoolers With High Reliability	287
	11.7	Cryocooler Design Using Rare-Earth Elements as Regenerator Materials	288
	11.8	Cryocooler With High-Pressure Ratio and Counterflow Design	288
		11.8.1 Thermodynamic Aspects of the Boreas Cycle	289
		11.8.2 Thermodynamic Efficiency Comparison for Various Coolers	290
		11.8.3 Advantages of High-Pressure Ratio Expansion	291
	11.9	Optimization of the Cooling Capacity of a Cryocooler	292
	11.10	Temperature Stability and Optimization of Mass Flow Rate	294
	Summary		298
	References		298

Index   **299**

Preface

The principal object of this book is to identify and describe applications of low-temperature superconductor/high-temperature superconductor (LTSC/HTSC) technologies to components and devices for use in radar, satellite communication, space, electronic warfare, medical diagnostic, electrical, and mechanical systems. Emphasis is on HTSC technology applications because of the minimum cost and complexity associated with cryocoolers at 77 K operations. However, integration of LTSC technology is necessary where high sensitivity, resolution, and efficiency are the principal requirements. Components and systems most suited for integration of HTSC and LTSC technologies have been clearly identified.

This book essentially provides adequate justification for effective integration of HTSC or LTSC technology in systems operating in various disciplines. The maturity of HTSC technology and the availability of HTSC materials in the near future will accelerate the development and production of superconducting radio astronomical systems, laser radars, sensitive MM-wave missile receivers, satellite communication systems, medical diagnostic equipment, diode-pumped solid-state lasers, space sensors, pollution monitoring sensors, infrared cameras, molecular spectroscopes, data storage devices, generators, motors, continuous wave infrared lasers, and Josephson junction devices with minimum weight, size, and power consumption. This book addresses issues related to the application of superconductivity technology to selected passive and active components and devices, such as A/D converters, filters, electrically small printed-circuit antennas, optical detectors, phase shifters, mixers, oscillators, eye-safe IR lasers, amplifiers, and high-power rf switches.

The book covers topics of interest to engineers, scientists, and researchers working in various fields. Both theory and applications are presented, with emphasis on the critical performance parameters of the superconducting components and systems. The book is written in a language accessible to undergraduate and graduate students who are interested in applications of superconducting technology. This book can be useful as a reference for physicists, research scientists, rf/microwave/analog engineers, project managers, educators, and clinical psychologists. In brief, it is of greatest value to those who wish to broaden their knowledge of the application of superconductor technology to hardware systems in various disciplines.

The author has made every attempt to provide a well-organized reference handbook using conventional nomenclature, a consistent set of symbols, and identical units. Mathematical expressions and derivations are provided wherever necessary. Performance parameters and experimental data on superconducting devices provided in this book have been taken from various references with due credit to the authors involved. The reference lists include the significant contributing sources, but are not exhaustive. The book has been designed to integrate research and development activities undertaken in the United States, Russia, Germany, England, France, Japan, and other advanced countries.

Chapter 1 deals with the theory of superconductivity formulated by Bardeen, Cooper, and Schrieffer in 1957 and the development of superconducting materials, including their unique features, namely, phase formation, crystalline structure, conduction, anisotropic properties of superconducting ceramic compounds, superconducting coherence, energy gap, grain boundaries, pinning energy, critical temperature, and lattice vibrations at cryogenic temperatures. Chapter 2 describes various forms of superconductors such as superconducting powder, bulk ceramic, wires, tapes, films, and ribbons. Critical current density, rf surface resistance, penetration depth, and Meissner effects are briefly discussed. Chapter 3 summarizes the electrical, mechanical, and thermal properties of superconducting substrates. HTSC/LTSC soft and hard substrates are briefly discussed, with particular emphasis on resonator unloaded quality factors (Qs) at cryogenic temperatures.

Chapter 4 deals with the application of superconducting technology to passive components and devices, such as filters, printed-circuit antennas, cavity resonators, digital circuits, delay lines, A/D converters (ADCs), detectors, and bolometers. Chapter 5 discusses the application of superconducting thin-film technology to active microwave components and devices for possible use in radar systems, space sensors, satellite communication, electronic warfare equipment, and electro-optical sensors. Projected performance parameters of cryogenically cooled phase shifters, electronically tunable filters, YIG-tuned filters, rf switches, power windows, and high-power ferrite circulators are specified.

Chapter 6 covers the application of superconducting technology to solid-state two-terminal and three-terminal devices, including PIN diodes, varactor diodes, Schottky barrier diodes, GUNN diodes, MESFETs, MODFETs,

HEMTs, p-HEMTs, and HBTs. Projected values of the performance parameters of solid-state devices, such as efficiency, noise figure, conversion loss, gain, and reliability as a function of cryogenic temperature are quoted wherever applicable. Chapter 7 deals with the application of superconducting technology to components and devices specifically used by radar, communication, electronic warfare, and space systems. Performance parameters for amplifiers, oscillators, receivers, antennas, multiplexing filters, A/D converters, and bandpass filters at cryogenic temperatures are summarized. Cryogenic operation not only provides improvement in noise figure, gain, reliability, and clutter rejection, but also offers significant reduction in system weight, size, and power consumption.

Chapter 8 describes the application of superconducting technology to electro-optical components and systems, which are widely used by both the military and civilian sector. Performance improvement in cryogenically cooled optical and infrared (IR) detectors, detector arrays, quantum-well laser diodes, IR cameras using focal planar arrays, forward looking IR systems, optical modulators, and scanning microscopes is briefly discussed. Chapter 9 deals with the application of LTSC and HTSC technologies to medical diagnostic techniques such as magnetic resonance imaging (MRI) and computer tomography (CT). Performance parameters of SQUID-based magnetometers and gradiometers are quoted as a function of cryogenic temperatures. Magnetometers and gradiometers using LTSC-SQUID devices permit biomagnetic measurements of human organs, with significantly improved resolution, accuracy, and reliability.

Chapter 10 discusses the application of superconducting technology to large turboalternators, motors, and power transmission lines. Published experimental data and results of the studies performed reveal that implementation of superconducting technology to these systems offers significant improvement in efficiency and considerable reduction in weight, size, and copper loss. The studies further indicate that low noise, high speed, enhanced efficiency, and improved reliability are possible only with levitating systems operating at lower cryogenic temperatures. Chapter 11 describes the design and performance requirements of cryocoolers. Issues related to regular maintenance and servicing of a cooling system to ensure safe, steady, and reliable operation of a superconducting system are addressed in detail. Input power requirements as a function of heat load and operating temperature are specified for various cryocoolers and microcoolers. Thermodynamic aspects of various cryocoolers are briefly discussed in terms of suction–pressure, efficiency, temperature stability, and mass flow rate.

The responsibility for the final form of the book, including errors of omission and commission, is mine. The scope of the book, the order of presentation of the material, and the sectional divisions within chapters are based on good engineering judgment and judicial choice. The timely completion of the work was not possible without editorial guidance and assistance from George

Telecki, Executive Editor of John Wiley & Sons, Inc. Finally, I wish to express my sincere thanks to my wife Urmila D. Jha for her patience and tolerance during the preparation of the manuscript.

A. R. JHA

Cerritos, CA
12 September 1997

Superconductor Technology

CHAPTER ONE

Phenomenology and Theory of Superconductivity

Superconductivity is a state of matter that is characterized by two distinct effects: zero resistance and diamagnetism, which means the explusion of magnetic fields. Superconductivity is a macroscopic phenomenon that involves amplitudes and phases asserted with the energy gap parameter. Interference and diffraction effects as well as the Josephson effect have been observed at cryogenic temperatures. These effects are employed in signal processors, rf circuits, EO components, data storage devices, and computer-related technology.

The universally accepted theory of superconductivity was formulated by Bardeen, Cooper, and Schrieffer in 1957, and is commonly known as the BCS theory. The BCS theory is the backbone of the superconductivity phenomenon, which states that if metallic mobile electrons interact efficiently with each other, then they exhibit

- Zero dc resistivity
- The Meissner effect, discovered by Meissner and Ochsenfeld in 1933
- A second-order phase transition to normal metallic state at a transition temperature
- Perfect diamagnetism in the presence of weak magnetic fields

The BCS theory offers an integral equation for the energy gap parameter (E_g) and another integral equation for the transition or critical temperature (T_c). These integral equations depend on the electron structure of the metal and on the degree of interaction between the electrons. The BCS model defines the energy gap parameter as

2 PHENOMENOLOGY AND THEORY OF SUPERCONDUCTIVITY

$$E_g = 1.76kT_c = 2hW_v \exp -\frac{1}{NV} \qquad (1.1)$$

where k = Boltzmann's constant
W_v = vibration frequency of the lattice
T_c = critical temperature (above which the material stops superconducting
N = number of available electronic states per unit energy in solid
V = voltage strength of attractive electron–electron interaction.

Equation 1.1 does not yield the most accurate results, because it represents an oversimplified model. The most accurate model based on the BCS theory was developed by Eliashberg and McMillion using precise experimental data on crystal lattice vibrations [1].

1.1 HISTORY OF SUPERCONDUCTING MATERIALS

Superconductivity was first observed in 1911 by the Dutch physics professor H. K. Onnes at the University of Leiden. Professor Onnes measured the resistance of several metals and was particularly surprised to see that the resistance of the mercury dropped to zero at 4.2 K. Low-temperature physics experiments in the following decades discovered other superconductors, including lead at 7.2 K, niobium at 8 K, niobium nitride at 15 K, and niobium germanium at 23 K.

1.1.1 History of Low-Temperature Superconductor (LTSC) Materials

The discovery of superconducting (SC) materials continued at moderate rate from roughly 1930 to 1980. During this period research and development activities focused primarily on LTSC materials and their applications. Practical application of the SC technique was first demonstrated in the early 1960s, when an AT & T Bell Laboratory scientist tested the high-current, high magnetic wires made from niobium–tin (Nb_3Sn) alloy. This test paved the way for the high magnetic field superconducting magnet industry and its applications. Almost simultaneously the Josephson effect was observed for the first time, which is responsible for the birth of Josephson junction electronic devices. The research on Josephson electronics grew quickly in the 1980s. In the 1970s and 1980s, LTSC solid-state device technology was applied to high-speed computer circuits, ultra-low-noise receivers, accelerators, and magnetic resonance imaging (MRI) sensors. LTSC materials and compounds are listed in chronological order of discovery in Table 1.1.

TABLE 1.1 Low-Temperature Superconductors

Material	Symbol	$T_c(K)$	Year Discovered
Mercury	Hg	4.2	1911
Lead	Pb	7.2	1913
Niobium	Nb	9.5	1933
Niobium nitride	NbN	16.0	1941
Vanadium silicon	V_3Si	17.1	1953
Niobium tin	Nb_3Sn	18.1	1960
Niobium aluminium germanium	NbAlGe	20.8	1969
Niobium germanium	Nb_3Ge	23.2	1973
Lanthanum barium copper oxide	LaBaCuO	30.0	1986
Barium potassium bismuth oxide	BaKBiO	30.4	1988
Lanthanum strontium copper oxide	LaSrCuO	39.0	1986

1.1.2 History of High-Temperature Superconductor (HTSC) Materials

The slow pace of scientific research on HTSC materials and compounds abruptly ended in late 1986, when Muller and Bednorz of the IBM Laboratory in Zurich (Switzerland) announced the discovery of a superconducting oxide at 30 K. In the spring of 1987, Paul Chu at the University of Houston announced the discovery of the yttrium barium copper oxide (YBaCuO or YBCO) HTSC compound with a transition temperature of 90 K. Research on HTSC materials grew rapidly and lead to the discovery of a bismuth compound (BiSrCaCuO or BSCCO) with a critical temperature of 110 K and thallium compound (TlBaCaCuO or TBCCO) with a critical temperature of 125 K. In May 1993, scientists in Switzerland discovered superconducting materials with critical temperatures in the range of 130 to 133 K. These superconducting materials are $HgBa_2Ca_2Cu_3O_{1+x}$ (with three CuO_2 layers per unit cell) and $HgBa_2CaCu_2O_6$ (with two CuO_2 layers). The superconducting structure comprises a defined sequence of the unit cells of these phases. The magnetic and resistivity measurements of the Swiss scientists confirm a maximum transition or critical temperature of 133 K.

New breakthroughs in superconducting materials and higher critical or transition temperatures were observed between 1988 and 1990. A wide variety of advanced or new superconducting compounds were investigated over the 1990–1992 period, with particular emphasis on thallium-based compounds. Recent aggressive research and development activities have pushed the transition temperatures well beyond 120 K. Specific details on the chronological discovery, number of Cu-O layers, and critical temperature for yttrium-based, bismuth-based, and thallium-based HTCS compounds are summarized in Table 1.2. All the important superconducting materials can be seen in Figure 1.1 in chronological order of discovery, with particular emphasis on transition temperature (T_c).

TABLE 1.2 High-Temperature Superconducting Compounds

Superconducting Compounds	Symbol	Transition Temperature (K)	Number of Cu-O Layers
$Tl_2Ba_2CuO_6$	TBCO (221)	90	1
$Tl_2Ba_2Ca_2Cu_3O_{10}$	TBCCO (2223)	125	3
$Tl_2Ba_2CaCu_2O_8$	TBCCO (2212)	110	2
$YBa_2Cu_4O_8$	YBCO (124)	81	2
$YBa_2Cu_3O_7$	YBCO (123)	95	2
$Bi_2Sr_2CaCu_2O_8$	BSCCO (2212)	90	2
$Bi_2Sr_2Ca_2Cu_3O_{10}$	BSCCO (2223)	110	3

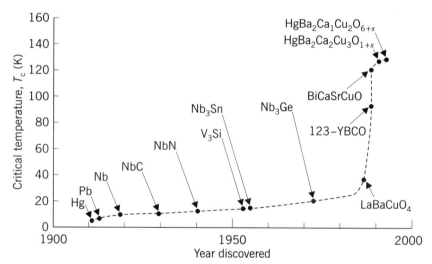

FIGURE 1.1 Chronological discovery of superconducting materials.

1.2 UNIQUE FEATURES OF SUPERCONDUCTORS

The unique features of superconducting materials involve anisotropic behavior, penetration depth, atomic orbitals, conduction layers, electrical doping, transport property, form, and microstructure. The performance of HTSC devices and sensors depends on the electrical and mechanical properties of the superconducting materials used in the fabrication of superconducting components. In some cases, conflict may occur between the electrical and mechanical properties of the HTSC materials or compounds. For example, the electrical performance of superconducting components or devices may be acceptable, but the brittleness of some ceramic compounds may pose a mechanical or structure problem. Therefore, careful selection of HTSC materials is necessary to meet specific electrical, mechanical, thermal, and structural requirements of the superconducting devices, components, or systems. In addition, integration of HTSC technology with MMIC devices involves special superconducting compounds, unique bonding methods, the right sputtering temperatures, appropriate oxygen pressures, and the latest fabrication procedures.

Relevant data on potential HTSC ceramic materials are summarized in Table 1.2. The HTSC superconductors shown in Table 1.2 are complex copper–oxide ceramic compounds. There are three principal subgroups: YBaCuO (YBCO), BiSrCaCuO (BSCCO), and TlBaCaCuO (TBCCO). These materials truly represent copper oxides. The various phases of these HTSC compounds depend on the number of CuO_2 layers and the valence of the chemical elements present in the compound (Table 1.3). Conduction results from the chemical reaction between the copper (Cu) and oxygen (O) atomic orbitals.

1.3 MICROSCOPIC STRUCTURES OF HTSC COMPOUNDS

HTSC materials have layered, oxygen-deficient perovskite structures in which fourfold planar-coordinated copper oxide (CuO) layers effectively form the conducting sheets. Macroscopic structural details of HTSC $YBa_2Cu_3O_7$ ceramic compound are shown in Figure 1.2, with particular emphasis on the Cu-O sheets. There are two Cu-O sheets in each unit cell, one on either side of the Y

TABLE 1.3 Valence of Elements Used in Superconducting Compounds

Element	Symbol	Valence
Aluminium	Al	3
Barium	Ba	2
Bismuth	Bi	2
Calcium	Ca	2
Copper	Cu	2,3
Lanthanum	La	3
Niobium	Nb	5
Oxygen	O	4
Strontium	Sr	2
Thallium	Tl	3
Tin	Sn	4
Yttrium	Y	3

Note. Nitrogen boils at 77 K, hydrogen boils at 20 K, helium boils at 4.2 K.

6 PHENOMENOLOGY AND THEORY OF SUPERCONDUCTIVITY

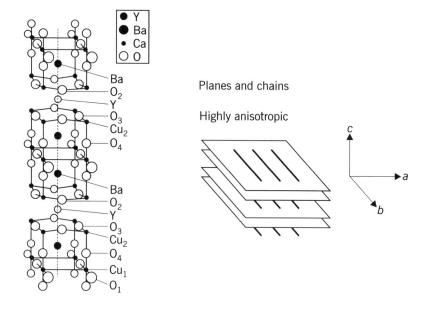

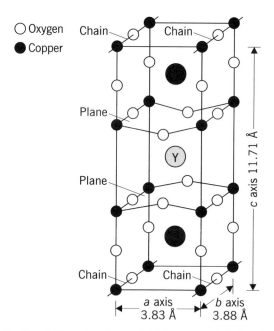

FIGURE 1.2 Crystalline structure of YBCO material showing chains, planes, and layers [2].

layer (yttrium layer). Conduction perpendicular to these sheets is very small because the oxygen linking the adjacent copper layers in that direction is almost absent, i.e., in the Y layer. This material can be treated as a two-dimensional superconductor. The conductivity in this material is very anisotropic and may even exhibit semiconducting-like characteristic along the c axis based on the transport properties. For the YBCO material there is a layer of copper-oxide chains along the b axis, in addition to the Cu-O sheets shown in Figure 1.2.

The other superconductor ceramic compounds, such as BSCCO and TBCCO, comprising copper oxides, differ primarily in the number of Cu-O sheets within the unit cell and the types of intervening layers separating them. However, among the HTSC materials listed in Table 1.2, the YBCO group is unique in having the Cu-O chains as one of the intervening layers (Figure 1.2). For a complete discussion on the chemistry and atomic structures of the HTSC materials, see reference [2].

1.4 PHASE FORMATIONS OF CRYSTALLINE STRUCTURES

It is important to understand the crystalline structure of (11) and (21) phases associated with the HTSC compounds. We will consider only the case of the BiSrCaCuO (BSCCO) or Bi(2212) ceramic compound in describing the phase formations. The crystalline structure of the phases of Bi(2212), (11), and (21) compounds is shown in Figure 1.3. Both (11) and (21) phases are orthorhombic structures, but they belong to different space groups. The Bi(2212) unit cell is basically a layered structure. Phase transformation occurs through melting processes in which CuO_2 layers lose oxygen (O) and are changed to CuO chains. The transition from the Bi(2212) to the (21) phase seems more straightforward than that from the Bi(2212) to the (11) phase. The (21) transformation involves rejection of more CuO chains from the (11) phase. There are 9 layers and 7 layers in one unit cell for the (11) and (21) phases, respectively (Figure 1.3). The phase formations in various HTSC compounds are dependent on the material composites, partial pressures, and sintering temperatures.

1.5 CONDUCTION

Conduction in HTSC materials is mostly due to the motion of the holes in the Cu-O sheets, but conduction due to carrier on some other layers cannot be ruled out. Conduction depends on the type and amount of dopants inserted in the compound. Electrical doping in the superconducting materials can be controlled by appropriate chemical substitution. In the case of $YBaCu_3O_7$, it is the oxygen content on the chain that undergoes variation not the Cu-O sheets. Therefore, electrical doping is controlled in each case by altering the superconducting material in those parts of the unit cell away from the Cu-O sheets

8 PHENOMENOLOGY AND THEORY OF SUPERCONDUCTIVITY

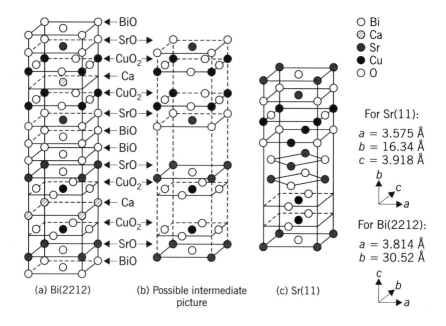

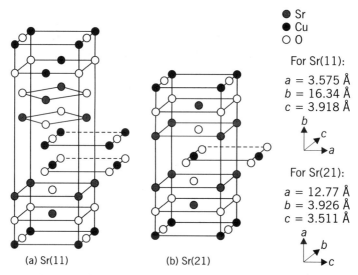

FIGURE 1.3 Phase transformation from BSCCO(2212) to Sr(11) and Sr(21) phases [4].

illustrated in Figure 1.2. In this respect the situation is like modulation doping in semiconductor superlattices. These HTSC ceramic compounds may exhibit insulating, conducting, or superconducting behavior, depending on the doping level. Such a wide range of electronic performance is possible due to simple chemical substitute with a given superconducting material.

1.6 ANISOTROPIC CHARACTERISTICS OF SUPERCONDUCTORS

The superconducting properties of the ceramic compounds are also anisotropic, which may affect some applications. The magnetic penetration depth in these materials is anisotropic. The magnetic penetration depth is about 1500 Å for the current flowing within the planes, but considerably higher for currents flowing perpendicular to the planes. The actual value of penetration depth is hard to determine, but is estimated to be in the range of 7500–9000 Å depending on the materials involved. The penetration depth for the current flowing perpendicular to the planes is about 7500 Å for the 123-YBCO material, but much larger for the 2212-TBCCO and 2212-BSCCO HTSC compounds. A material with deeper penetration is characterized as more isotropic.

Deep penetration for the current flowing perpendicular to the oxide layers may cause crossover, particularly in a-axis and c-axis films, where the current flow is forced to change direction and flow perpendicular to the layers. The change in direction causes large inductance and low critical current levels. Special deposition procedures and film growth techniques are required to overcome the problems created by the anisotropic penetration depths of HTSC materials.

1.7 SUPERCONDUCTING COHERENCE

The superconducting coherence lengths of superconducting materials are not completely known, but they are considered isotropic [3]. Estimates vary from 10 to 20 Å within the conducting plane for conventional superconducting materials. However, the coherence length in the YBCO varies over 2 to 4 Å, which is on the order of separation between the Cu-O layers. The coherence length is even shorter (less than 2 Å) for the higher anisotropic materials such as BSCCO or TBCCO. Short coherence lengths will put demands on HTSC film growth quality and chemical uniformity near the surfaces or interfaces, which can affect the function and performance of the device. Extremely short coherence lengths can cause serious problems at grain boundaries.

1.8 SUPERCONDUCTING ENERGY GAP

Existence of a superconducting energy gap and tunneling effects in the anisotropic materials have been observed over the last few years. The superconducting energy gap in anisotropic materials exists between ab planes (Figure 1.2), and the gap along the c axis is smallest, which seriously constrains the ability of the film configurations to meet specific performance requirements. Any constraint put on the preparation of thin films requires tough quality control inspection, which drives the cost of the device too high.

1.9 GRAIN BOUNDARIES

HTSC materials can be obtained in bulk ceramic, thin-film, and single-crystal form, each with different grain boundaries. The schematic microstructure of high-temperature complex oxide superconductors in bulk ceramic, thin-film, and single-crystal form is shown in Figure 1.4. The highly anisotropic nature of the HTSC materials seriously impacts their grain boundaries with poor conductivity. The straight lines within these grains represent the edges of the conducting *ab* planes. The circles indicate individual grains within the anisotropic material. The clear areas in Figure 1.4 represent the surface of the *ab* planes with the *c* axis normal to them. It is evident from Figure 1.4 that the ceramic superconducting material consists of grains with random orientations. Special processing techniques are required to align the grains necessary for high-quality material. However, the degree of grain alignment and extent of

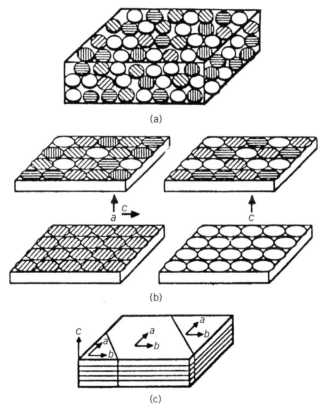

FIGURE 1.4 Schematic microstructure of HTSC oxides in (a) ceramic, (b) thin-film, and (c) single-crystal materials. Straight lines represent the edges of the *ab* planes, circles represent individual grains within the material, and clear areas represent the surfaces of the *ab* planes.

grain boundaries vary from polycrystalline thin film to a single-crystal film to bulk sample. Specific details on the grain orientation and grain boundary of a bulk sample and a thin film are shown in Figure 1.5.

Conduction paths can be considerably diluted by severely disoriented grains, unless the material is single crystal. Good transport and strong superconducting properties are necessary to achieve a highly orientated grain structure in a superconducting material. YBCO material is called orthorhombic due to the presence of alternate directions of the a and b axes. Twin boundaries exist within the individual grains in a 123-YBCO thin film.

The coupling strength is strictly related to connectivity, rather than to to the orientation of the grain boundaries in the coupled sample. High crystal coherence can be achieved by the facet structure at the grain boundaries of thin film. Low-energy grain boundaries generally are strongly coupled and are not easily penetrated by the magnetic flux lines. Strong coupling applies to high-angle boundaries, whereas weak coupling exists at low angle grain boundaries. The

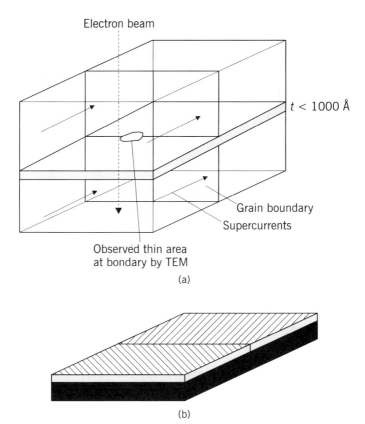

FIGURE 1.5 Grain boundaries of high-temperature superconducting (a) bulk sample and (b) thin film.

12 PHENOMENOLOGY AND THEORY OF SUPERCONDUCTIVITY

critical current density will be severely reduced only when the grain boundaries are penetrated by the flux lines. Specific details on the weakly coupled and strongly coupled regions in the grain boundaries can be seen in Figure 1.6. Grain boundaries are treated as planar defects in the bulk superconducting materials. However, the grain boundaries in thin films may be considered line defects and the orientation will have greater effect on the transport critical current density. The relationship between the transport critical current density (J_c) and the coupling mechanism of the grain boundaries (GB) is evident in Figure 1.7.

As shown in Figure 1.6 the transport current density at high field is proportional to the area of strongly coupled regions at grain boundaries. Experimental data on transport current density versus applied magnetic field (H) on the YBCO processed material can be seen in Figure 1.7 [4]. The

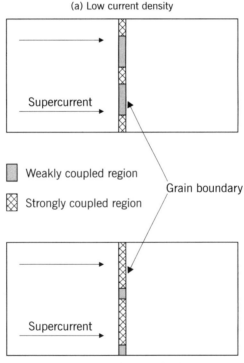

FIGURE 1.6 Schematics showing (a) the low current density and (b) the high current density regions at grain boundaries. Transport current density (J_c) exhibits weak-link behavior generally in sintered polycrystalline samples due to second-phase coating at the boundary area and misorientation of grains. Most of the Josephson junction weak links can be decoupled by the applied magnetic field. At higher sintering temperatures, some strongly coupled regions can be formed due to liquid-phase sintering.

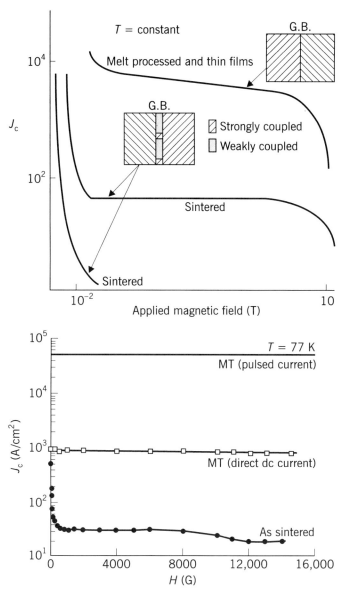

FIGURE 1.7 Critical current densities as a function of applied magnetic field under various test conditions.

coupling strength of the grain boundaries is a complex issue in which several factors must be considered. Grain orientation effect must be considered in conjunction with other important structural and superconducting properties, such as crystal coherence, mechanical connectivity, stoichiometry, critical temperature (T_c), and superconducting related magnetic field.

14 PHENOMENOLOGY AND THEORY OF SUPERCONDUCTIVITY

The conduction paths can be considerably diluted due to severe disoriented grains unless the material is single crystal. A highly oriented grain structure in a superconducting material is necessary to achieve good transport and strong superconducting properties. Since YBCO is widely used as a superconducting material, our discussion is primarily focused on this material. YBCO is called orthorhombic because the a and b axes have alternate directions. In addition, twin boundaries exist within the individual grains of the thin film in this particular superconducting material.

1.10 PINNING ENERGY

The properties of the vortex lattice and the associated flux-pinning behavior of the HTSC materials described in Table 1.2 are different from those for the low-temperature conventional superconducting materials shown in Table 1.1. The pinning energy is dependent on grain alignment and grain boundary, but the grain alignment and grain boundary vary from polycrystalline-film to single-crystal film to bulk sample. The new superconducting materials have short coherence lengths, which lead to small pinning energies with respect to thermal energies, thereby leading to a very large magnetic flux creeping rates. For a comprehensive discussion on the pinning energy effects, see reference [5].

1.11 CRITICAL TEMPERATURE

Critical temperature (Table 1.4) is the most important characteristic of a superconducting material. The transition temperature and the sintering temperature depend on the substrate and the superconducting material used for

TABLE 1.4 Critical Temperature(T_c), Sintering Temperature(T_s), and Critical Density for in Situ Growth of YBCO Film Deposited on Various Substrates

Substrate	T_s (°C)	T_c (K)	Current Density (A/cm$_2$) at	
			4.2 K	77 K
SrTiO$_3$	830	94	6×10^6	5×10^5
SrTiO$_3$	600	92	2×10^6	3×10^6
SrTiO$_3$	400	85	5×10^6	8×10^6
MgO	720	87	8×10^5	10^6
MgO	600	80	10^6	10^5
MgO	650	82	10^6	10^5

Note. These properties are subject to growth condition over a wide range of sputtering temperature, oxygen pressure, and the thermodynamic stability line of 123-YBCO toward decomposition into Y$_2$BaCuO$_5$, BaCuO$_2$, and Cu$_2$O. The roles of equilibrium thermodynamics, kinetics, and the chemical reactivity of the species need further investigation.

the film. The new HTSC compounds discovered over the 1986–1988 period have a large transition temperature, critical current density, critical magnetic field, and energy gap. Milestones of research and development activities directed toward HTSC ceramic oxides are summarized in Table 1.5. They contain mostly four- or five-component copper oxides (Cu-O), except for the copper-free bismuth oxides. These materials with high critical temperatures have a very low critical current and exhibit an isotope effect and a rich variety of solid-state phases. Some of the phases are antiferromagnetic and some are insulating. The high-temperature, copper-based superconducting phases are very anisotropic with strong two-dimensional or one-dimensional characteristics. These HTSC compounds are, in all respects, traditional superconductors and can be manufactured with minimum cost and complexity. Antiferromagnetism and superconductivity are known to coexist in some cases, but magnetic moments tend to destroy, not enhance the superconductivity state.

TlBaCaCuO (TBCCO) has zero resistance at a transition temperature or critical temperature of 125 K, but thallium is so poisonous that it may not be suitable for some applications. Special handling and a rigid operational environment may be required if thallium is used. According to IBM research scientist Ed Angler, the critical temperature can be varied by controlling the number of planes in the thallium material. The two-plane 2-2-1-2 ($Tl_2Ba_2CaCu_2O_8$) phase has a T_c of 108 K, the three-plane 2-2-2-3 ($Tl_2Ba_2Ca_2Cu_3O_7$) phase has a T_c of 125 K, and a third phase (1-2-2-3) with a combination of two and three planes has a T_c of 118 K. In $YBa_2Cu_3O_7$ material, which is generally designated as 123-YBCO, the superconducting phase of 1-2-3 contains the atoms in the ratio of one yttrium to two barium to three copper.

1.12 LATTICE VIBRATIONS

The lattice vibrations of HTSC ceramic oxide compounds are unusually soft and are primarily responsible for high transition or critical temperatures. Generally, lattice vibrations play a minor role in the superconducting properties of the ceramic oxides. These oxides suffer from material problems. They can be brittle, hard to handle, and hard to machine and have unstable phases and low critical currents. Innovative material processing techniques are being developed to overcome these problems. The new HTSC materials can be produced in bulk, thin-film, or single-crystal form.

1.13 IMPACT OF MAGIC ANGLE ON TRANSITION TEMPERATURE

Transition temperature may be influenced by a magic-angle effect [6], whereby the alignment of strong magnetic fields along certain angles or orientations causes sudden and dramatic changes in the electrical conductivity of the super-

TABLE 1.5 Milestones of Research and Development Activities in HTSC Ceramic Oxides

Month Announced	Milestone	Institution	Leading Researchers
April 1986	Lanthanum barium copper oxide superconducting at temperature of 30 K	IBM-Zurich	Alex Muller Georg Bednorz
December 1986	IBM's April findings confirmed and zero resistance temperature of LaBaCuO raised	University of Tokyo	Shoji Tanaka Koichi Kitazawa
	LaBaCuO (under pressure) superconducting at 40 K	University of Houston	Paul Chu
January 1987	Lanthanum strontium copper oxide super-conducting at 36 K	AT&T Bell Laboratories, Murray Hill, N.J.	Robert Cava
	LaBaCuO superconducting at 70 K	Chinese Academy of Sciences Institute of Physics, Beijing	Z. X. Zhao
February 1987	Yttrium barium copper oxide superconducting at 95 K	University of Houston with University of Alabama-Huntsville	Paul Chu with Mau-Kuen Wu
May 1987	Critical current densities in excess of 10^5 A/cm^2 achieved in YBaCuO	IBM-Yorktown Heights	Praveen Chaudhari
October 1987	No data distortion during transmission in ceramic superconductor at 100 KHz	Cornell University University of Rochester	Robert Buhrman Gerard Mourou
November 1987	Critical current density of 7000 A/cm^2 achieved in bulk YBaCuO	AT&T Bell Laboratories, Murray Hill, N.J.	Sungho Jin
January 1988	Bismuth strontium calcium copper oxide superconducting at 106 K (zero resistance at 85 K)	National Research Institute of Metals, Tsukuba, Japan	Hiroshi Maeda
February 1988	Thallium barium calcium copper oxide fully superconducting at 106 K	University of Arkansas	Allen Hermann Z. Z. Sheng
March 1988	ThBaCaCuO superconducting at 125 K	IBM-Almaden	S. Parkin

conductor. This possibility was predicted by the Russian scientist A. G. Lebed in 1989. New findings, including the experimental investigation being carried out at the State University of New York at Buffalo, seem to prove the existence of a magic angle. The results may lead to higher transition temperatures (greater than 145 K) for organic and copper-oxide superconductors in near future. It is clear from the limited data available that a tremendous increase in conductivity is possible at the magic angle, leading to a higher transition temperature.

1.14 INTEGRATION OF SUPERCONDUCTING TECHNOLOGY WITH OTHER TECHNOLOGIES

The Nobel-prize-winning discovery in 1987 of cuprate-based superconducting materials, which led to rapid development of materials with transition temperatures well above of liquid nitrogen, sparked worldwide research and development activities. Laboratory experiments demonstrate that this new technology is sufficiently robust to survive space and military applications. Substantial reduction in losses, weight, size, and power consumption have been realized by employing the superconducting technology. Dc motors using superconducting technology can deliver output greater than 60 times that of conventional motors. The ac synchronous motors with field windings consisting of HTSC coils have shown remarkable reduction in weight, size, and losses. Using HTSC coils could result in considerable energy savings in large industrial motors, where small increases in efficiency can add up quickly.

Recently a low-loss, compact, microwave filter has been developed for application in satellite communication equipment, which indicates the integration of HTSC technology into monolithic microwave integrated circuit (MMIC) technology. Superconducting technology has been integrated in infrared (IR) detector technology, with remarkable improvement in the detector responsivity. Low-temperature superconductor technology led to the development of magnetic resonance imaging (MRI), which has revolutionized medical technology. Superconducting technology has been applied to acoustic transducer development under a Small Business Innovation Research (SBIR) contract from the U.S. Navy. The HTSC acoustic transducer employs a magnetostrictive element and a pair of magnet coils made of HTSC wire and a Stirling-cycle cryorefrigerator to maintain the device temperature at 50 K using no liquid nitrogen. The acoustic transducer is the most critical component for underwater sonar applications that convert electrical power into low-frequency acoustic power with high efficiency using the magnetostriction property. The HTSC acoustic transducer can also be used in environmental applications, seismic tomography, ocean-floor mapping, global warming indicators, and ocean currents monitors.

Recent advances in superconducting electronic technology are leading to the day when Josephson junctions start replacing conventional semiconductor devices in digital processors and systems. Superconducting technology can be used in digital applications because of a unique property: the quantization of magnetic flux in a superconducting loop. This property is widely used in the design and development of superconducting logic circuits and superconducting flash A/D converters. High resolution, ultrawide bandwidth, large dynamic range, fast response, minimum size, and low power consumption are the major advantages of these devices. Superconductivity also plays a key role in electrical engineering, optical engineering, rotating power equipment, high-power lifting magnets, propulsion systems, magnetically levitated trains, power generation, power transmission, and power distribution. We discuss the commerical and military applications of superconductivity devices and technology throughout this book.

SUMMARY

Superconductivity is a state of matter characterized by two distinct properties: zero resistance and diamagnetism. The BCS theory is the backbone of the superconductivity phenomenon. Superconductivity was first observed in 1911 by the Duth physics professor H. K. Onnes. Most LTSC materials were discovered from 1911 to 1960. The discovery of HTSC compounds accelerated from 1980 to 1988. Thallium ceramic compound (TlBaCaCuO) had the highest known transition temperature (125 K) until 1993, when Swiss scientists discovered two new HTSC materials: $HgBa_2Ca_2Cu_3O_{1+x}$ (with three CuO_2 layers) and $HgBa_2CaCu_2O_{6+x}$ (with two CuO_2 layers) with transition temperatures of 133 and 130 K, respectively.

Superconducting materials are unique in anisotropic behavior, penetration depth, conduction layers, microstructure, phase formation, coherence length, grain boundaries [7], pinning energy, lattice vibrations, and critical temperature. Superconductivity technology has made possible the development of MMIC, acoustic transducers, IR detectors, satellite communications and medical imaging with high resolution.

REFERENCES

1. R. D. Parks, ed. *Superconductivity* (2 vols.). New York: Dekker Publishing Co, 1969, p. 682.
2. A. W. Sleight. Chemistry of high-temperature superconductors. *Science* 242:1519–1527 (1988).
3. D. M. Ginsburg, ed. *Physical Properties of High-Temperature Superconductors*. Singapore: World Scientific, 1989.
4. R. B. Poeppel, editor-in-chief. *Appl. Superconductivity* (1/2):59 (1993).

5. D. R. Nelson. Vortex entanglement of high-temperature superconductors. *Phys. Rev. Lett.* 6:1973–1976 (1988).
6. Rodman Publishing. *Superconductor Industry*, 2:36 (Summer 1993).
7. R. B. Poeppel et al. Transport critical currents and grain boundary weaklink. *Appl. Superconductivity* (1/2):67–69 (1993).

CHAPTER TWO

Superconductor Forms and Their Critical Microwave Properties

High-temperature superconductor (HTSC) materials are found in various forms, such as powder, single-crystal, polycrystalline, and bulk. Superconductors are fabricated in four distinct forms: wire, tape, ribbon, and film. Bulk materials are suitable when machining for specific structures and shapes is involved. Wires are used for low-power magnets, antennas, medium torque motors, and compact electrical generators. Superconducting tapes are most useful where large lengths of material are required and must meet stringent weight and size requirements. Thin films are widely employed in microwave devices, optical detectors, and microelectronic circuits. HTSC thick films are best for high-power microwave cavity applications. Ribbons are most suitable for spaceborne and airborne superconducting devices, where implementation of monolithic microwave integrated circuit (MMIC) technology is highly desirable. Recent data on HTSC forms and methods are summarized in Table 2.1 [1].

2.1 GENERAL PROPERTIES AND APPLICATIONS

Each superconducting form has unique electrical, magnetic, and mechanical characteristics. Specific forms are required to meet performance specifications and packaging requirements. For example, low-power analog devices and superconducting quantum interference device (SQUID) sensors are fabricated from superconducting thin films, whereas high-power microwave components require HTSC thick films. High-power magnets, motors, and generators generally require long tape or filament. Some special systems, such as particle

TABLE 2.1 Recent Data on Various Superconducting Forms as a Function of Frequency [1]

Characterization Method	Measurement Frequency (GHz)	Materials	R_{dc}
Cylindrical cavity	100	YBCO film on $SrTiO_3$ substrate	0 @ 90 K
Stripline resonator	0.5–18	YBCO film on $YZrO_2$ substrate	0 @ 72 K
Disk resonator	10	Bulk YBCO material	0 @ 92 K
Cylindrical cavity	9.8	Bulk YBCO ceramic	0 @ 92 K
Cylindrical Nb cavity with HTSC	6.0	Crystal platelet	0 @ 89 K
Half-wavelength resonator	0.2–0.4	YBCO powder	—

R_{dc} is the dc component of the resistance.

accelerators, magnetic resonance imaging (MRI) systems, and large-capacity storage devices, may require a combination of superconductor forms. As far as the critical current capability is concerned, YBCO films can have values greater than 1×10^6 A/cm^2, YBCO wire greater than 1280 A/cm^2 and BSCCO bulk sample greater than 175 A/cm^2. Transport critical current density, flux expulsion, flux-pinning mechanisms, phase diagrams, and mechanical properties must be given serious consideration, regardless of the form selected.

2.2 SUPERCONDUCTING POWDER

A superconductor in powder form normally does not play a key role in the fabrication of a device or component. Nonoxide ceramic powders are the most popular. Aluminium nitride powder is widely used in the preparation of other superconducting forms. It is generally produced using a carbon thermal reduction process, which offers a high-purity product with low particle size. Particle size is the most important characteristic of a superconducting powder. Typical properties of aluminium nitride powder are summarized in Table 2.2.

2.3 SUPERCONDUCTING BULK MATERIALS

Properties of bulk ceramic materials can be varied by changing the sintering parameters and liquid processing methods. Transport critical current density, sintering parameters, electron microscopic properties, and grain boundaries are of major concern in bulk materials. Structural differences may exist between strongly coupled and weakly coupled grain boundaries. A high-angle grain boundary is of primary importance and can be characterized by three distinct

TABLE 2.2 Properties of Aluminum Nitride Powder

Aluminum	65–66%
Hydrogen	<0.1%
Nitrogen	33%
Oxygen	0.5–1.0%
Carbon	0.4–1.5%
Particle size	0.2–1 µm
Surface area	6–12 m^2/g

Note. Aluminium nitride ceramic powder is produced using a carbothermal reduction process, which provides a high-purity product with low particle size.

types: (1) amorphous layer, (2) transition structure, and (3) coincidental site lattice boundary. Amorphous-type grain boundaries are formed in sintered bulk samples. A bulk sample is formed through a solid-state diffusion process over a temperature range of 900–1000°C. The sintering process through solid-state diffusion is rather slow and the grains are randomly oriented.

Sintering parameters play a key role in shaping the superconducting properties of a sample. Sintering below 900°C for 12 h does not increase the current density of the samples beyond 70% of the theoretical value. Samples with 100% theoretical density can be obtained by sintering at a temperature of 1000°C for 2 h. The coupling strength of the grain boundaries in a bulk material is dependent on the structural and superconducting properties, which include crystal coherence, mechanical connectivity, stoichiometry, transition temperature, and critical magnetic field. Even though the latest bulk material is rather coarse and porous by microwave standards, its surface resistance is much lower than that of gold at 77 K. The important properties of superconducting bulk material are summarized in Table 2.3.

2.3.1 Optimization of Processing Conditions for Bulk Ceramics

Processing conditions vary among bulk superconductor materials. The processing technique and conditions used in case of bulk YBCO samples can attribute the flux pinning dominated transport current density to strongly coupled, high-angle grain boundaries. Substitution of elements with different ionic radii and different bonding character can significantly improve zero-resistance temperature.

Processing techniques used by the nuclear scientists at the University of Illinois [2] revealed that lithium (Li) doping in the superconducting compound of $Bi_2Sr_2CaCu_2O_x$ (BSCCO-2212) reduces the melting temperature be 140°C, increases the transition temperature (T_c) more than 10%, and improves flux pinning under high magnetic field environments (greater than 20 T). However, for a Li-doped BSCCO-2212 material, the T_c behavior is extremely sensitive to sintering temperature compared to that of undoped BSCCO-2212 material.

TABLE 2.3 Properties of HTSC Ceramic Bulk Materials

Property	Powder	Bulk (1-mm rod)	Polymer Composite
Superconducting transition temperature (K)	92	92	[a]
Critical current density (A/cm^2)	—	1000	—
Flexural strength (MPa)	—	200	60
Young's modulus (GPa)	—	180	20
Work of fracture (J/m^2)	—	20	500–10^5
Critical stress intensity factor (MPa/m^2)	—	1.1	Not linear elastic
Thermal conductivity (W/m K^{-1})	—	2.67	0.65–1.23
Thermal diffusivity (23°C) (cm^2/s)	—	0.012	0.003–0.006
Coeff. of thermal expansion (25–500°C) (μm/mK^{-1})	—	11.5	12–25
Specific heat (J/g K^{-1})	—	0.431	0.42–0.55
Density (g/cm^3)	6.38	5.4–6.38	4–5
Density (% of theoretical)	100	85–100	—
Surface area (m^2/g)	4–5	—	—
Purity after firing (%)	>98	>98	—

[a] Exhibits Meissner effect below T_c.
Note. M, mega; G, giga; Pa, pascal (1 psi = 6890 Pa).

Sintering a pellet twice not only increases the transition temperature, but also improves the microstructure.

The Li-doped superconductor is extremely sensitive to both procesing conditions and parameters. Because of greater sensitivity, the processing conditions and parameters must be optimized and controlled to meet specific superconducting device performance requirements.

2.4 SUPERCONDUCTING WIRES

Superconducting wires are available in diameters ranging from 1 to 10 mm. HTSC wires can be used in low-frequency antennas, low-power motors, and low-power electrical generators. Special processing techniques are used to make superconducting wires. The melt process is generally used in fabricating the silver-sheathed (Ag-sheathed) YBCO wires and (Bi Pb)$_2$Sr$_2$CaCu$_2$O$_y$ wires, known as BPSCCO-2212 wires. A high-T_c phase formation decomposition–recovery process (PFDR) involving a short-period partial wet process is used to recover the phase in the core of the wire, with subsequent annealing. Decomposition of superconducting phases during the melting process increases the critical current density (J_c) due to significant improvement in texturing and densification.

A high current density and improved current density-magnetic field dependence can be achieved through the high T_c BPSCCO-2223 formation using the

PFDR process. The current density varies from $40,000$ A/cm^2 at 77 K and 0 tesla (T) magnetic field to 25,000 A/cm^2 at 0.1 T and 9000 A/cm^2 at 1 T in the PFDL-processed superconducting tapes and wires. High mass density, excellent grain alignment, uniform distribution with small impurity particles, and large grain sizes have been achieved through the PFDR process, which offers significant improvement in current density and the electromagnetic properties of the wires. Powder processing techniques and thermomechanical processes prior to fabrication must be optimized to achieve high mechanical integrity of the wire. Brittleness is the most serious problem associated with the superconducting wires because of the silver core. Further research and development activities on HTSC wires are required to overcome this problem.

Mechanical properties in terms of tensile stress, flexural strength, and torsional strength for superconducting wires must be investigated as function of operating temperature, core diameter, core material, and turn radius. Limited information available on mechanical properties of the wires indicates that the bismuth lead-based wires are stronger than YBCO-based wires. Thallium-based wires appear to be stronger than both the bismuth lead-based and YBCO-based wires under similar operating conditions.

2.5 SUPERCONDUCTING TAPES

Superconducting tapes are widely used in applications where large lengths of materials are required to meet the high-power performance and stringent weight and size requirements. Silver-clad tapes of BPSCCO-2223 material can be produced from polyphase powders, which reactively sinter in situ to form the BPSCCO-2223. Optimum annealing temperatures and unique thermomechanical methods are necessary to achieve the highest critical current density in the tapes. Reactive sintering of polyphase tapes is extremely sensitive to annealing temperature. There is a narrow range of annealing temperature, and the optimum annealing temperature must be used to allow maximum transformation without decomposition of the 2223 phase.

Continued intermittent processing after initial grain-growth annealing is necessary to advance the chemical reaction and to promote grain texturing. The highest current density has been seen in BPSCCO tapes made from polyphase powder annealed for 350 h at 845°C with three intermittent pressing cycles (Figure 2.1). Continued thermomechanical processing (Figure 2.2) leads to a remarkable increase in current density as a result of improved grain texturing. Completion of the annealing process between 830 and 853°C indicates that maximum transformation of 2212 to 2223 due to endothermic reaction takes place over this narrow temperature range. A suitable number of thermomechanical cycles and intermediate annealing times with minimum deformation are essential to achieve high current density.

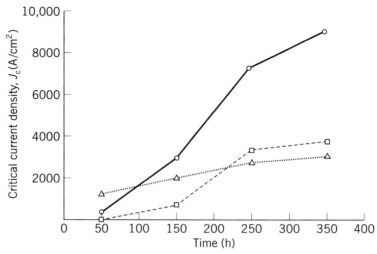

FIGURE 2.1 Critical current density as a function of annealing time at 845°C. Tapes are pressed after 250 h (○), 150 h (□), and 50 h (△) [3].

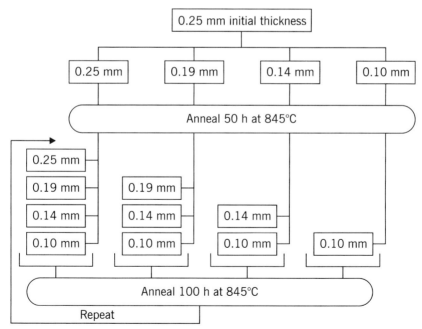

FIGURE 2.2 Flow chart of mechanical processing of silver-clad (Bi, Pb)$_2$Sr$_2$Ca$_2$Cu$_3$O$_x$ tapes [3].

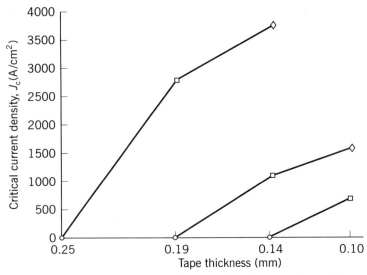

FIGURE 2.3 Critical current density as a function of initial annealing thickness during three cycles: ○, 50 h; □, 150 h; ◇, 250 h [3].

According to W. Lo and B. A. Glowacki [3], the critical current density tends to increase with decreasing tape thickness (Figure 2.3), because texturing decreases from the Ag-BPSCCO interface into the ceramic core [4].

2.6 SUPERCONDUCTING FILMS

Superconducting films have opened up new possibilities for passive and active microwave and optical components, namely, filters, delay lines, microstrip patch antennas, power combining circuits, solid-state devices, kinetic induction phase shifters, low-noise MM-wave amplifiers, MRI sensors, SQUID devices, A/D converters, and optical detectors. Superconducting films less than 1 μm thick are called thin films, whereas films greater than 1 μm are classified as thick films. Low-noise devices and passive components generally use thin films, while high-power components mostly employ thick films to achieve good power-handling capability.

2.6.1 Superconducting Thin Films

HTSC thin films of YBCO, BSCCO, BPSCCO, and TBCCO have been successfully deposited on several substrates, including $LaAlO_3$, MgO, $SrTiO_3$, and Al_2O_3. Appropriate processing techniques are used to achieve high quality thin films. The film processing technique is optimized for each superconducting film and its substrate. Surface resistance, insertion loss, quality factor, and critical current density characteristics are important to a superconducting component designer. Some of the properties of thin film are briefly described.

2.6.1.1 Insertion Loss The predominant loss factors in HTSC films are largely due to material impurities and dielectric losses in the substrate medium. The insertion loss in the metallic medium is significantly less than that in the dielectric medium. The insertion loss in metallic films is strictly due to surface resistance (R_s), which is a function of operating frequency and cryogenic temperature. An HTSC thin film offers loss performance that is at least two orders of magnitude better than that of copper films, particularly over the 1 to 10-GHz frequency spectrum. Calculated values of radio-frequency (rf) surface resistance for various thin films as a function of temperature and frequency are shown in Figure 2.4 [5]. The overall attenuation or insertion loss in a microstrip circuit is the sum of losses in the strip and in the substrate. The overall insertion loss in a microstrip transmission line using a thin film of YBCO deposited on an yttria-stabilized zirconia (YSZ) substrate can be seen in Figure 2.5 as a function of frequency and temperature [6].

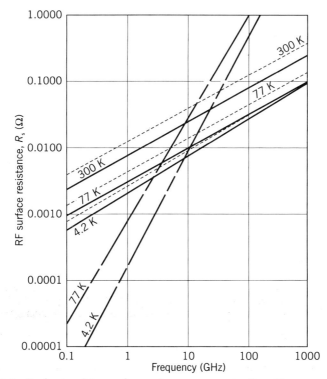

FIGURE 2.4 Surface resistance for various superconducting films as a function of temperature and frequency: ---, gold film; —, copper film; – – –, YBCA thin film.

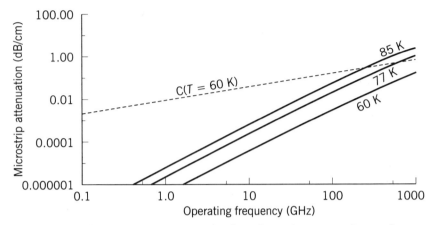

FIGURE 2.5 High-temperature superconductive microstrip attentuation performance as a function of cryogenic temperature and operating frequency: —, YBCO material on yttria-stabilized zirconia (YSZ, $e_r = 27$ at 10 GHz); ---, copper at 60 K on YSZ substrate (thickness = 0.020 inches = 500 µm).

It is evident from this figure that HTSC thin films of YBCO offer the lowest insertion loss at 60 K, even up to 1000 GHz, compared to copper film.

Dielectric losses occur in granular substrate materials due to their porosity, whereas the metallic losses are primarily due to chemical nonstoichiometric impurities that are by-products of some film preparation processes. In bulk granular HTSC materials, uniform grain sizes can minimize microwave losses due to an exponentially enhanced scattering phenomenon present in the medium.

HTSC films must be deposited on a low-loss dielectric substrate, such as lanthanum aluminate ($LaAlO_3$), which had dielectric constant of about 24.5 and a loss tangent of 0.00003 at X-band frequencies. The films can be supplied with single-sided or double-sided structure deposited on a gold ground plane, as shown in Figure 2.6. The TBCCO thin films are patterned by a wet etch

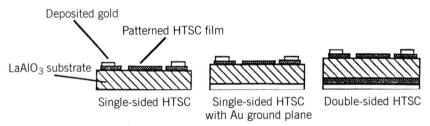

FIGURE 2.6 Cross-section views of single-sided and double-sided microstrip circuits using HTSC thin films. For specific details on the performance capabilities, see Reference [7].

process [7]. Gold is deposited directly on the HTSC surface wherever an rf contact is required. A single-sided HTSC microstrip configuration can be used when only one conductor surface is required, as in a coplanar waveguide. The double-sided HTSC microstrip structure is most useful in applications where highest unloaded Q (Q_{un}) and lowest insertion loss are the principal requirements. Substrate losses are discussed in detail in Chapter 3.

2.6.1.2 Quality Factors YBCO epitaxially thin films grown on single-crystalline $LaAlO_3$ substrate using the pulsed excimer laser ablation technique offer a current density exceeding $10E^7 A/cm^2$ at cryogenic temperatures, in addition to high unloaded Qs. TBCCO thin films on $LaAlO_3$ substrate offer high unloaded Qs (Q_{un}) and ultralow loss at frequencies greater than 10 GHz. Calculated values of unloaded Qs for thin films of YBCO on MgO and $LaAlO_3$ substrates are summarized in Table 2.4.

2.6.1.3 Transition Temperature of Thin Films Thin films of various superconducting materials have been developed with high transition temperatures. The transition temperature limit can be improved by adding more layers of CuO_2 in the HTSC compounds. Finite interaction is necessary between the CuO_2 layers in the superconducting ceramic compounds to achieve high values of T_c. This concept has improved transition temperatures in thin films of bismuth-based and thallium-based superconducting materials. Calculated values of transition temperature values for the bismuth-based (BSCCO) superconducting compound as a function of CuO_2 layers are shown in Figure 2.7. In thallium-based films, the improvement in T_c is slightly higher as a function of the number of CuO_2 layers.

Thallium-based (TBCCO) thin films are prepared using an off-axis rf magnetron sputtering method followed by postannealing in thallium oxide vapor. The thallium-based thin films are highly c-axis oriented with minimum phases.

TABLE 2.4 Unloaded Qs of YBCO Thin Films on MgO and $LaAlO_3$ Substrates as a Function of Frequency and Temperature

Frequency (MHz)	MgO ($e_r = 10.0$)		$LaAlO_3$ ($e_r = 24.5$)	
	77 K	4.2 K	77 K	4.2 K
100	15,600	39,600	46,000	72,000
1,000	7,800	19,800	23,100	36,150
3,000	2,600	6,600	7,700	12,000
6,000	1,300	3,300	3,850	6,000
9,000	870	2,200	2,580	4,000
18,000	435	1,100	1,290	2,200

Note. The highest unloaded Qs offer minimum insertion loss at cryogenic temperatures.

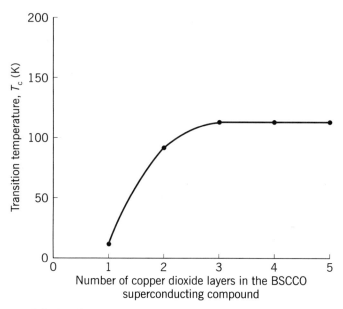

FIGURE 2.7 Calculated values of transition temperature for BSCCO superconductor as a function of layers. Finite interaction is necessary between the CuO_2 layers to achieve a high transition temperature.

TBCCO films less than 0.5 µm thick have demonstrated the lowest rf surface resistance and superior environmental stability at 10 GHz and 77 K.

The surface resistance values of YBCO thin films on MgO substrate are higher than those for TBCCO thin films because of higher transition temperature of the thallium-based superconductor. The surface resistance increases with the increase in operating temperature and input power drive, which is evident from the curves shown in Figure 2.8 for YBCO films on MgO substrate [8]. Furthermore, the quality factor decreases with the increase in drive power (Figure 2.9a). The resonance frequency shift with increase in the normalized temperature is small at low drive power levels. However, the frequency shift increases rapidly at higher operating temperature and higher power levels exceeding 0 dBm (Figure 2.9b). This resonance frequency shift is due to a kinetic inductance effect, which is highly visible as the operating temperature approaches the transition temperature of the superconducting material. Kinetic inductance contributes an effective permeability, which reduces the velocity of propagation.

2.6.2 Superconducting Thick Films

Superconducting films greater than 1 µm thick are classified as thick films. Thick films are useful where high-power operation over a long duration is the principal requirement. High-power microwave cavities have been developed using thick films of YBCO and TBCCO materials on sapphire substrates

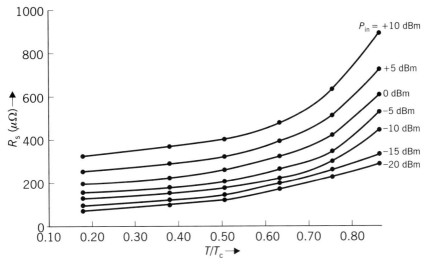

FIGURE 2.8 Microwave surface resistance of YBCO thin film as a function of input power and normalized temperature [8]: T/T_c = normalized temperature, T = operating temperature, T_c = critical temperature of the HTSC material, R_s = surface resistance.

to meet high-power and high critical current density requirements. HTSC thick films are generally magnetron-sputtered onto an oriented silver alloy substrate to provide high electrical conductivity. High-power microwave cavities require thick films to achieve the desired accelerating gradients.

BSCCO thick films coated on a silver plate and YBCO thick films coated on a silver wire have been developed for use in devices requiring large superconducting surfaces [9]. YBCO thick films coated on silver wires have shown rf properties similar to those of the bulk ceramic and have the advantages of being fabricated into samples of large surface areas and complex shapes with minimum cost and complexity.

TBCCO thick films have been developed on large surfaces using a magnetron-sputtering technique onto oriented silver alloy substrate. These films exhibit the highest degree of c-axis (optical axis) texturing, which indicates the weakest dependence of surface resistance on rf drive power. They also exhibit the sharpest transition to the superconducting state based on measured data obtained at higher microwave frequencies [10].

Thallium-based thick films on sapphire substrates have demonstrated the highest values of unloaded Q as a function of operating temperature and circulating power. These films have also exhibited the lowest values of surface resistance as a function of current density and operating temperatures. The characteristics of TBCCO thick films on sapphire are shown in Figure 2.10 [11]. TBCCO films have unloaded Qs greater than $10E^7$ at 4.2 K and 1000 W of circulating power. The performance of TBCCO films remains unmatched to this date.

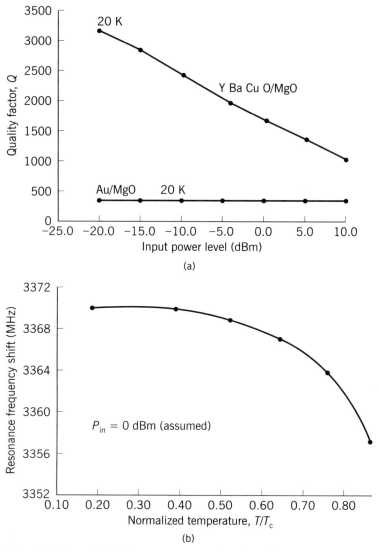

FIGURE 2.9 Variation of YBCO parameters as a function of input power and normalized temperature: (a) quality factor as a function of input power; (b) frequency shift as a function of temperature.

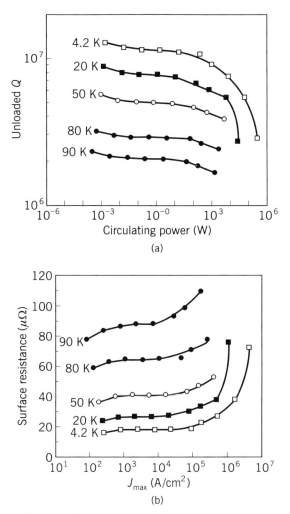

FIGURE 2.10 (a) Unloaded Q as a function of power and temperature at 5552 MHz [10, 11], and (b) surface resistance as a function of critical current density and temperature for TBCCO films on sapphire.

2.7 CRITICAL MICROWAVE PROPERTIES

The most important microwave properties of superconductors are the critical current density, radio-frequency surface resistance, and critical magnetic field. Based on the critical magnetic field, the new HTSCs have been divided into two categories. The type 1 superconductor has only one critical magnetic field value, below which current flowing over the surface cancels the magnetic field inside the superconducting material. The type 2 superconductor has a second higher critical magnetic field capable of trapping and magnetic flux inside the material. Surface resistance and critical current density depend on the superconductor form factors, substrate properties, cryogenic temperature, and rf drive power level.

2.7.1 Critical Current Density

Critical current density is important in applications where high-power handling capability is the principal requirement. Critical current densities as a function of cryogenic temperature for the YBCO films are summarized in Table 2.5. YBCO films enjoy the highest values of current density from 4.2 to 77 K. The YBCO in situ films have higher values than the thick films. The critical current densities for the BSCCO postannealed films are comparable to those of the YBCO postannealed films, despite their higher critical temperatures. The critical current density of the c-axis-oriented TBCCO thick film and a polycrystalline YBCO thin film have the lowest values of current density. Polycrystalline YBCO thin films suffer from the adverse effects of poor grain orientations. Performance of the TBCCO thick films can be further improved with good grain orientation. Thick film of TBCCO are most suitable for use in high-power microwave cavities.

2.7.2 Radio-Frequency Surface Resistance

Microwave circuit loss is proportional to rf surface resistance, which varies as the square root of the operating frequency for a normal conductor. However, in a superconductor, it is proportional to the square of the frequency and

TABLE 2.5 Critical Temperature (T_c), Sintering Temperature (T_s), and Critical Density for in Situ Growth of YBCO Film Deposited on Various Substrates

			Current Density (A/Cm²)	
Substrate	T_s (°C)	T_c (K)	4.2 K	77 K
SrTiO$_3$	830	94	6×10^6	5×10^5
SrTiO$_3$	600	92	2×10^6	3×10^6
SrTiO$_3$	400	85	5×10^6	8×10^6
MgO	720	87	8×10^5	10^6
MgO	600	80	10^6	10^5
MgO	650	82	10^6	10^5

Note. These properties are subject to growth conditions over a wide range of sputtering temperatures, oxygen pressure, and the thermodynamic stability line of 123-YBCO toward decomposition into Y$_2$BaCuO$_5$, BaCuO$_2$, and Cu$_2$O. The relation between equilibrium thermodynamics, kinetics, and the chemical reactivity of the species needs further investigation.

directly proportional to the operating temperature. This means higher losses are expected in a superconductor at higher frequencies under noncryogenic operations. The superconductor theoretically has zero loss at a temperature well below its critical temperature. In brief, the surface resistance is a function of operating temperature, frequency, and microwave properties of superconducting and substrate materials.

This behavior can be explained by the two-fluid model theory, which postulates that both superconducting electrons and normal electrons exist below the critical temperature. As the operating temperature is reduced, the number of superconducting electrons increases until the temperature reaches zero. At this temperature all electrons theoretically are superconducting electrons. For zero dc resistance, it is necessary only to have a few superconducting electrons. However, in case of rf surface resistance the situation is different. The oscillating motion of the superconducting electrons induces a reactive voltage due to the inertia of the electrons, which is responsible for rf losses. The faster the electrons move back and forth, the higher will be the rf losses in a superconductor. The speed of the electron movement is proportional to the operating frequency, which indicates an increase in surface resistance with frequency at a given temperature.

Radio frequency losses can be reduced if the operating temperature is maintained well below the critical temperature of the superconducting material, because at the critical temperature the number of superconducting electrons start to dominate the number of normal or regular electrons. This means that the operating temperature must be kept well below the critical temperature to keep losses low compared to copper. A superconducting material with a low critical temperature is thus highly desirable for low-noise and low-loss microwave devices.

The overall rf resistance of a superconductor can be expressed as

$$R_{ORF} = R(T)_{res} + R(f, T) \tag{2.1}$$

where $R(T)_{res}$ = temperature-dependent residual resistance
$R(f, T)$ = resistance function of frequency and temperature
f and T = operating frequency and temperature

The last parameter in the above equation is defined as

$$R(f, T) = Cf^n \exp{-\frac{E_g}{kT}} \tag{2.2}$$

where C = constant, n = 1 to 2, E_g = energy gap, T = absolute temperature (K), and k = Boltzman's constant.

The value of the residual resistance for superconductors is very small at operating temperatures well below the critical temperature of the superconducting material. The magnitude of the temperature-frequency-dependent resistance component can be calculated using the relevant material parameters. The net value of this component depends on the film surface conditions, grain sizes, grain orientations, transition temperature, operating frequency, operating temperature, and the exponent n. For a well-developed niobium (Nb) superconductor, $R(f, T)$ at 4.2 K is about three orders of magnitude lower than it is at room temperature (300 K) at 10 GHz and about two orders of magnitude lower than it is at 100 GHz. The liquid helium temperature of 4.2 K is roughly one-half of the critical temperature for the Nb superconductor, which is in good agreement with the rule of thumb that superconductors must be operated at about 50% of their transition temperatures if optimum performance is the principal objective.

Based on this rule thumb, the YBCO superconductor with T_c of about 91 K must be operated at around 45 K to achieve optimum device performance. High values of superconductor resistance are strictly due to granularity and anisotropic properties of the ceramic superconducting compounds. In general, lower surface resistance has been observed with superconducting thin films of higher quality, regardless of the materials. Thin film performance appears to be much closer to that of an ideal single crystal. This indicates that surface resistance and, thus, microwave loss are lower in thin films than in bulk ceramics or thick films.

2.7.3 Penetration Depth

Penetration depth in a superconductor medium is an important performance parameter. The penetration depth varies from one superconductor to another and is a function of London penetration depth at zero temperature. Thorough knowledge of the skin-depth parameter is essential to appreciate the importance of penetration depth in a superconductor. The skin depth (δ) is inversely proportional to the square root of the frequency for a normal conductor, which indicates that at higher microwave frequencies the rf current is concentrated in a thin outside layer of the conductor. For a high-quality superconducting thin film, the penetration depth is about 2000 Å, while for a gold-plated normal conductor the skin depth is about 20,000 Å, which is ten times the penetration depth in a superconductor.

London penetration depth (λ_0) is generally used in mathematical analysis for superconducting transmission lines, which indicates how far a magnetic field can penetrate into a superconducting material. London penetration depth is much smaller than magnetic penetration depth and is independent of frequency. Typical values of London penetration depth vary from 1000 to 3000 Å, depending on the type of superconductor. Even at very low microwave frequencies, the thickness of a superconducting film can be a few micrometers and still can exhibit low rf losses, which is not possible with normal conductors. Computed values of rf surface resistance as a function of operating temperature and frequency were shown in Figure 2.4.

2.7.4 Meissner Effect

In superconducting materials, the magnetic flux lines are completely expelled and there is a force pushing superconductors away from the magnetic fields. This effect is called the Meissner effect and was discovered by Meissner and Ochsenfeld in 1933. The Meissner effect indicates that the nonsynthesized sample of Bi(2212) has a critical temperature of 90 K (Figure 2.11), which is a typical value for the 2212 cuprates. The Meissner effect for the BiPb(2223) sample with critical temperature greater than 110 K is also shown in Figure 2.11. The lower inflection point occurs due to the effects of oxygen and lead on the 2212 sample. The 2223 sample phase is slightly affected by treatment at 500°C and 165 bar pressure. The experimental data support the conclusion that the 2223 phase is still stable at 500°C and 165 bar pressure. Three supercon-

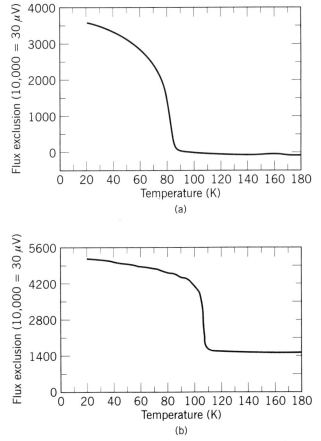

FIGURE 2.11 Flux exclusion as a function of temperature for (a) Bi(2212) and (b) BiPb(2223) films.

ducting phases exist for the Bi-based and BiPb-based superconducting compounds depending on the number of $Cu-O_2$ layers [9].

SUMMARY

Superconducting materials are found in powder, bulk, single-crystal, and polycrystalline forms, but they are generally manufactured in four distinct forms: wires, tapes, ribbons, and films. Bulk materials are best used when machining for specific structures and shapes is involved. Superconducting wires are generally used in superconducting low-power magnets, antennas, medium torque motors, and compact electrical generators. Ribbons and films are widely used when integration of MMIC technology with superconductor technology is highly desirable. Superconducting tapes are employed when large lengths of superconducting material is required. Thin films are generally used in passive rf and microwave devices, microelectronic circuits, and optical detectors. Superconducting thick films are widely used in high-power microwave cavities, high-power lifting magnets, and populsion systems.

Optimization of processing techniques and conditions is required for each superconducting form to achive state-of-the art performance. In superconducting bulk ceramics, substitution of elements with different ionic radii is necessary to improve the zero-resistance temperature. A proven process is highly essential to obtain superconducting wires with high critical density and improved current density–magnetic field dependence. Further research is necessary to overcome the problem of brittleness associated with superconducting wires. Optimum annealing temperatures and unique thermomechanical methods are required to achive high current density in HTSC tapes without decomposition of the essential superconducting phase of the material. Transition temperature, critical current density, quality factor, surface resistance, penetration depth, and Meissner effect are the most critical characteristics of superconducting materials.

REFERENCES

1. A. Fathy et al. Microwave characteristics and characterization of high T_c superconductors. *Microwave J.* 92 (October 1988).
2. S. Wu et al. Optimization of processing conditions for bulk ceramic lithium-doped BSCCO. *Appl. Superconductivity* (1/2): 93–99 (1993).
3. W. Lo and B. A. Glowacki. *Superconductor Science Technical* 4:S361 (1991).
4. D. Y. Kaufman et al. Thermomechanical processing of reactively sintered Ag-clad (Bi, Pb)$_2$Sr$_2$Ca$_2$Cu$_3$O$_x$ tapes. *Appl. Superconductivity* (1/2): 81–91 (1993).
5. A. R. Jha. Application of HTSC technology in MM-wave components and circuits. *Proc. 3d Asia–Pacific Microwave Conference*, Tokyo (Japan), September 1990, pp. 607–610.

6. A. R. Jha. Application of high-temperature superconducting thin films in MM-wave circuits. *Proc. 15th International Conference on Infrared and MM-Waves*, Orlando (Florida), December 1990, pp. 163–165.
7. Superconductor Technologies, Inc., 460-F Ward Drive, Santa Barbara, CA 93111-2310. Product sheet, May 1990.
8. R. Pinto et al. Power and temperature dependence of Q factor of a double-sided thin-film YBCO microstrip resonator. *Appl. Superconductivity* (1/2):1–6 (1993).
9. H. Muller et al. *Proc. Superconductivity Conference*, San Francisco (California), August 1988, pp. 21–23.
10. D. W. Cooke et al. Absorption of high microwave power by large-area Tl-based superconducting films on metallic substrates. *IEEE Trans. MTT* 39:1539–1544 (1991).
11. Z. Y. Shen et al. High-T_c superconductor–sapphire microwave resonator with extremely high Q values up to 90 K. *1992 IEEE MTT-S Digest*, pp. 193–196.

CHAPTER THREE

Superconducting Substrate Materials

Electrical, mechanical, and physical properties of high-temperature superconducting (HTSC) substrates are important to performance requirements and operating conditions. Substrates must meet the electrical as well as the mechanical properties, if operation under severe conditions is the principal requirement. Dimensional stability, anistropy effect, coefficient of contraction, dissipation factor (or loss tangent) as a function of temperature and frequency, and dielectric constant (or permittivity) as a function of temperature and frequency are the most critical substrate parameters. Suitable substrate materials must be identified for deposition of HTSC films of YBCO, BSCCO, TBCCO, and other superconducting films.

In this chapter, critical, properties of soft, hard, and metallic substrates are discussed. HTSC substrates for passive, low-power, and high-power rf components are identified. Applications requiring hard substrates are briefly mentioned. Properties of metallic substrates, which are considered most ideal in fabrication of high-power microwave cavities using thick films, are discussed. Interface problems between the superconducting films and substrates materials are identified wherever possible [1].

3.1 CLASSIFICATION OF HTSC DIELECTRIC SUBSTRATES

Various dielectric substrate materials are used in microwave, optical, microelectronic, and MM-wave components and circuits. Dielectric substrates can be classified as soft and hard. Pure Teflon, CuFlon, ROHACELL foams, microfiber PTFE, and woven PTFE are called soft substrates. Sapphire (Al_2O_3), alumina ceramic (99.6% pure SiO_2), fused silica or quartz (100% SiO_2), strontium titanate (SrTiO$_3$), magnesium oxide (MgO), zirconium oxide (ZrO$_2$), and lanthanum aluminate (LaAlO$_3$) belong to the hard substrate category.

3.2 SOFT DIELECTRIC SUBSTRATES

Soft dielectric substrates include pure Teflon ($e_r = 2.20$), CuFlon ($e_r = 2.15$), RT/DUROID 5880 ($e_r = 2.22$), RT/DUROID 5870 ($e_r = 2.35$), RT/DUROID 6002 ($e_r = 2.94$), and RT/DUROID 6010 ($e_r = 10.5$). The RT/DUROID materials may be classified as Teflon fiberglass laminates, which are generally referred as to PTFE (polytetrafluoroethylene)/glass laminates. There are two types of PTFE/glass laminates: woven and unwoven (microfiber). Both meet the performance requirements of soft substrates.

3.2.1 Electrical Properties

Critical electrical and rf properties of HTSC soft substrates include dielectric constant or permittivity (e_r) as a function of frequency and temperature, dissipation factor or loss tangent as a function of frequency and temperature, and anisotropy, which is defined as the ratio of dielectric constant of the medium along xy plane to that in the z direction (e_{rxy}/e_{rz}). The difference in the value of e_r along xy and z directions occurs when a fill material such as ceramic or glass is added to achieve high dimensional stability. The addition of fiberglass or ceramic to the basic substrate medium will increase both the permittivity and the loss tangent of the finished product. However, glass loading must give way to ceramic loading for high dielectric constant substrate. Dimensional stability is achieved at the expense of high anisotropy. The closer the anisotropy to unity, the more electrically uniform will be the material.

Typical values of anisotropy for various HTSC soft dielectric substrates are summarized in Table 3.1 with specific comment on dimensional stability of the material. Anisotropy varies from 1.00 for pure Teflon to 1.01 for CuFlon, 1.025 for microfiber PTFE, and 1.16 for a woven PTFE material. The lowest values of both the dielectric constant and dissipation factor are necessary to achieve minimum dielectric losses at microwave and MM-wave frequencies.

3.2.2 Relative Dielectric Constants and Loss Tangents

Comprehensive studies on the critical dielectric properties of potential soft substrates were performed, including Roger Corporation's RT/Duroid 5880, 6002, and 6010 materials as a function of frequency and temperature. Our studies indicate that the relative dielectric constant increases for most of the soft substrates along all the three axes as the operating temperature is reduced. Figure 3.1 shows that rapid increase in the normalized value of the permittivity for the RT/Duroid 6010 occurs as the operating temperature is reduced below 77 K. Figure 3.2 shows a rapid decrease in the value of dissipation factor for the same

TABLE 3.1 Typical Anisotropic Values for Various HTSC Soft Substrates [1]

Substrate	e_{xy}/e_z	Dimensional Stability
Teflon (PTFE) ($e_r = 2.10$)	1.000	Excellent
Microfiber PTFE ($e_r = 2.20$)	1.025	Good
Microfiber PTFE ($e_r = 2.35$)	1.041	Fair
Woven PTFE ($e_r = 2.17$)	1.092	Poor
Woven PTFE ($e_r = 2.45$)	1.160	Worst
High dielectric Duroid ($e_r = 10.5$) with $t = 0.025$ in.	1.0201	Good
High dielectric Duroid ($e_r = 10.5$) with $t = 0.050$ in.	1.060	Fair
High dielectric Duroid ($e_r = 10.5$) with $t = 0.075$ in.	1.180	Worst

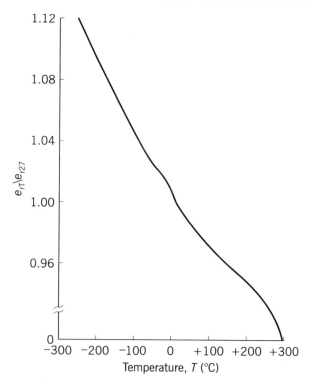

FIGURE 3.1 Impact of temperature on the normalized permittivity of RT/Duroid 6010 at 10 GHz: relative dielectric constant at 27°C = 10.80; e_{r27} = relative permittivity at 27°C (300 K); e_{rT} = relative permittivity at absolute temperature T (K).

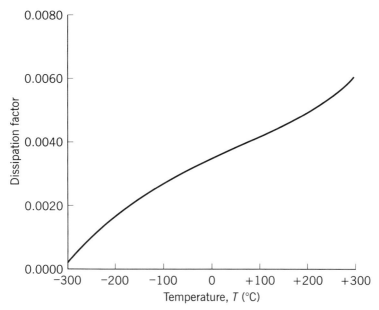

FIGURE 3.2 Loss tangent or dissipation factor for RT/Duroid 6010 substrate as a function of temperature at 10 GHz.

material as the temperature of the substrate is reduced. The values of the dissipation factor for RT/Duroid 6010 in Figure 3.2 were calculated by subtracting the loss factor of copper, which is assumed to be 0.00079 at 300 K and 0.00001 at 4.2 K, from the overall loss factor of the microstrip using copper thin film on this substrate [2]. The author has examined the characteristics of RT/Duroid 6010, which is a ceramic-PTFE composite substrate best suited for microwave and microelectronic circuit applications requiring high values of permittivity. This substrate offers excellent performances at cryogenic temperatures, precision dielectric constant control, low moisture absorption, excellent dimensional stability, and nearly isotropic electrical properties at low temperatures.

The values of the dielectric constant for RT/Duroid 6002 as a function of temperature and frequency are summarized in Figure 3.3. The measured values of the dissipation factor for this substrate as a function of temperature are also shown. The dielectric constant for this substrate increases with the decrease in temperature, which is different from other soft substrates.

The dispersion effects in HTSC substrates at microwave frequencies have also been investigated. Frequency dispersion effects in thin soft substrates (not exceeding 0.010 inches) are negligible at higher frequencies, because of minimum variation in the permittivity of the substrate, particularly at temperatures below 77 K. Effective dielectric constant values [3] as a function of temperature are shown in Figure 3.4 for two soft substrates, RT/Duroid 6010 and RT/Duroid 5880. The effective dielectric constant varies by 10% for the

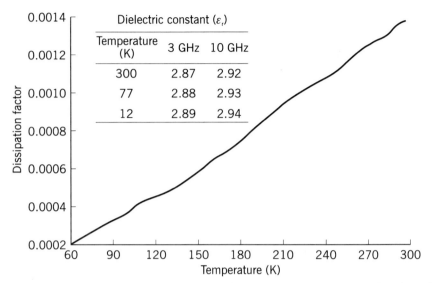

FIGURE 3.3 Measured values of dissipation factor and permittivity as a function of temperature for the soft substrate RT/Duroid 6002.

6010 material compared to less than 1% over the same temperature range for the 5880 material. This indicates that the frequency dispersion effect is negligible in the 5880 substrate material despite its large thickness of 0.031 inches. If the thickness for the 6010 substrate material is reduced to 0.005 inches, the dispersion effect is hardly noticeable. In brief, a soft substrate with low permittivity ($e_r < 2.20$) and thickness not exceeding 0.010 inches will be free from frequency dispersion for this particular material over the entire temperature range from 300 to 0 K. Extrapolation of these values down to 0 K is fully justified, because there is no sufficient evidence of thermal transition effects over this temperature range. It is evident from Figure 3.5 that the tensile modulus [4] and, hence, the tensile strength will experience a linear gain in magnitude as the temperature is reduced, particularly, over the temperature range from 100 to 0 K, with a maximum value occurring at 0 K in both directions. The increase in the tensile modulus of pure Teflon is relatively small compared to that for the microfiber PTFE material 5880 operating at 77 K.

Studies performed on foam-based substrates reveal the following:

1. The permittivity or the dielectric constant for the foam-based substrate increases as the temperature is reduced from 300 to 77 K, as with other soft substrates.
2. The loss trangent value increases as the temperature is reduced to 77 K, which is an opposite reaction to that seen in other soft substrates.

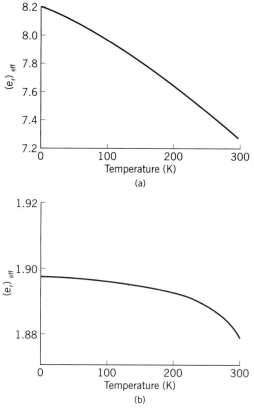

FIGURE 3.4 Measured values of effective dielectric constant for various substrates as a function of temperature: (a) RT/Duroid 6010 (0.025 inch) ($e_r = 10.5$ at 10 GHz and 300 K; (b) RT/Duroid 5880 (0.031 inch) ($e_r = 2.20$ at 10 GHz and 300 K). e_r, relative dielectric constant; e_{reff}, effective dielectric constant, which is highly dependent on the substrate thickness and type. These curves are based on measurements made at Mullard Radio Astronomy Observatory, University of Cambridge, Cambridge, UK, 5 September 1983 [3].

3. The tensile and compressive strengths improve by 15 and 36% respectively, as the temperature is reduced to 77 K.

3.3 HARD DIELECTRIC SUBSTRATES

Sapphire, alumina ceramic, fused quartz, fused silica, strontium titanate, magnesium oxide, zirconium oxide, and lanthanum aluminate belong to hard substrate category. Hard substrates such as magnesium oxide (MgO), alumina ceramic (Al_2O_3), 99.6 to 99.9% pure), lanthanum aluminate ($LaAlO_3$), and

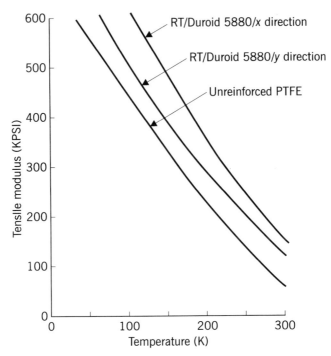

FIGURE 3.5 Tensile modulus for RT/Duroid 5880 as a function of temperature.

sapphire (Al_2O_3) have been widely used in the development of superconducting passive and active microwave components and circuits. Alumina ceramic is available in all shapes and sizes with minimum cost. This particular substrate is widely used where moderate performance with minimum cost is desired. The dielectric constant is affected by the preferred orientation of the crystals and the density of the material, regardless of the substrate.

3.3.1 Critical Electrical Properties

3.3.1.1 Alumina Ceramic Alumina material is widely used in superconducting components and circuits, because of its minimum cost and abundant supply. The most popular alumina ceramics are available with 99.6 to 99.9% purity. Fine-grained alumina ceramic has a more random crystalline orientation than course-grained material. The uniform electrical properties depend on the grain size and random orientation, but random orientation is strictly dependent on the accuracy of the process control followed during the manufacturing phase of the material.

The adhesion HTSC films on alumina substrates depends on the density of the grain boundary areas. Fine grain size produces a high-density superconducting substrate material that reduces the microwave scattering effect, which is primarily caused by a coarse grain structure. Excessive scattering will result in higher microwave losses in the substrate medium even at cryogenic temperatures. An alumina substrate with a fine grain structure is required to achieve optimum permittivity and minimum loss tangent at low temperature and over a wide microwave frequency spectrum.

Substrate flatness is important for achieving a uniform dielectric constant over a wide frequency and temperature range. Surface finish is another important requirement for a high-performance substrate. Polished substrates offer a better rf performance level than fired or rough substrates. The dielectric constant for an alumina substrate (99.6% pure) varies from 9.90 ± 0.02 to 9.91 ± 0.02 over a frequency range of 0.1–25 GHz and a temperature range of 77–300 K. The dissipation factor varies from 0.00010 at 300 K to 0.00008 at 77 K over a frequency range of 1–50 GHz. The surface finish for this material is better than 2.5 microinches and the material is available with thickness as low as 0.005 inches. Both the fine surface and thin substrate are essential for HTSC MM-wave devices and circuits.

3.3.1.2 Quartz/Fused Silica (100% SiO_2)

This substrate has a dielectric constant of 3.78 over the entire frequency range of 0.1–25 GHz and over the temperature range of 77–300 K. Its dielectric constant or permittivity is somewhat higher than that of woven PTFE/glass substrate material, but it has the lowest coefficient of thermal contraction, minimum anisotropy, and excellent dimensional stability. This material offers ultralow dielectric losses even at MM-wave frequencies. It is highly recommended for microwave active and passive devices, where low insertion loss and high dimensional stability at MM-wave frequencies are the principal requirements.

Fused silica possesses superior mechanical properties, which makes it most attractive for HTSC components or circuits operating under severe mechanical conditions. This substrate has demonstrated a tensile strength better than 8560 psi and a compressive strength more than 170,000 psi at room temperature (300 K). It has a tensile modulus of 10×10^6 psi, a shear modulus of 4.4×10^6 psi, and a modulus of rupture better than 5500 psi, all at an operating temperature of 77 K. Projected values of tensile modulus and shear modulus for fused silica substrate as a function of temperature are plotted in Figure 3.6. Both parameters decrease as the temperature is reduced.

Fused silica substrate offers the lowest coefficients of thermal expansion (CTE) and contraction (CTC). The CTE of the material is $-0.6\,\mu m/°C$ at 77 K (Figure 3.7). A linear drop in CTC occurs over the temperature range of 0–77 K. A low coefficient is highly desirable, because the substrate is generally attached to a case and also might have been soldered or epoxied to it. To prevent the solder or epoxy joints from cracking or deformations at cryogenic temperatures, the values of CTC for the substrate and case materials must be not only low, but also as close as possible to maintain high performance, reliability, and mechanical integrity.

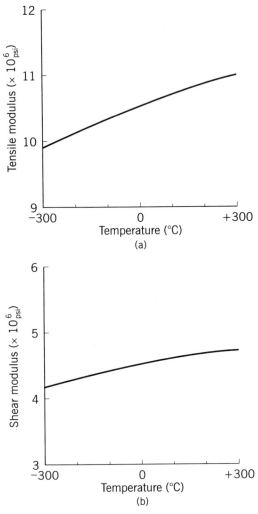

FIGURE 3.6 (a) Tensile modulus and (b) shear modulus as a function of temperature for fused silica.

3.3.1.3 Sapphire Al_2O_3 Sapphire is a hard substrate material with a room temperature (300 K) permittivity of about 10.8. Since sapphire is not an isotropic material, its dielectric constant may vary from 9.3 to 11.0, depending on the orientation. This material offers unloaded Q (Q_{un}) as high as 10 million and a dissipation factor of 0.00015 at 10 GHz and 77 K. Its dissipation factor is close to that of alumina, quartz, and fused silica. Furthermore, this substrate has demonstrated much higher quality factors than alumina ceramic, magnesium oxide, and lanthanum aluminate at cryogenic tempera-

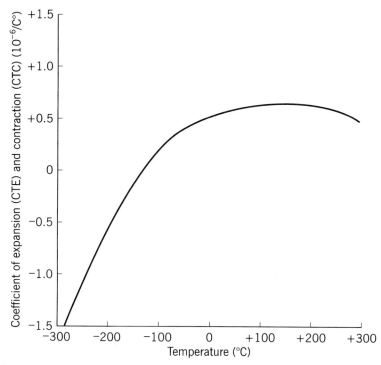

FIGURE 3.7 Coefficient of expansion/contraction for fused silica as a function of temperature.

tures. Typical values of quality factors for various HTSC conductors and substrates are summarized in Table 3.2. The overall quality factor values for the HTSC microstrip line will be lower than those of the substrate material. Theoretical quality factor values for various HTSC substrates as a function of frequency can be seen in Figure 3.8 [5].

TABLE 3.2 Quality Factors for Various Conductor and Substrate Materials at 10 GHz

Conductor	(Q_c)
Copper (300 K)	500
Copper (77 K)	1,500
HTSC at 1/10 copper (77 K)	15,000
HTSC at 1/100 copper (4.2 K)	150,000

Substrate	(Q_d)
Alumina	10,000
MgO	10 to 50,000
LaAlO$_3$	10 to 100,000
Sapphire	10,000,000

50 SUPERCONDUCTING SUBSTRATE MATERIALS

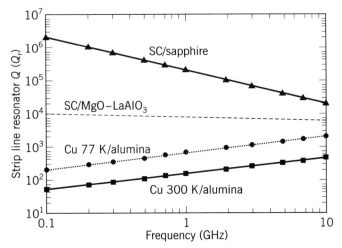

FIGURE 3.8 Theoretical strip line quality factor at 77 K for various HTSC substrates [5]. $1/Q_r = 1/Q_c + 1/Q_d$, where Q_r = resonator Q, Q_c = conductor Q, Q_d = dielectric Q, Q = quality factor.

Thin films of niobium ($T_c = 8$ K), YBCO ($T_c = 91$ K), and TBCCO ($T_c = 110$ K) when deposited on a sapphire substrate have demonstrated excellent electrical characteristics at cryogenic temperatures. Calculated values of insertion loss and effective dielectric constant at 4.2 K as a function of frequency for a superconducting microstrip line involving niobium thin film on a sapphire substrate are shown in Figure 3.9. The insertion loss (IL) has been calculated using the following empirical formula:

$$IL = (1.59)(10E - 5)f^2 \text{ dB/mm} \tag{3.3}$$

where f is the frequency (GHz).

The effective dielectric constant for the above substrate at 4.2 K varies linearly over the frequency spectrum up to 16 GHz. Thin films of niobium on sapphire substrates are capable of yielding excellent microwave performance over a 1- to 25-GHz frequency range.

3.3.2 HTSC Substrate Requirements at MM-Wave Frequencies

Substrate materials for superconducting MM-wave circuit applications require ultralow values of permittivity, loss tangent, and anisotropy at cryogenic temperatures. Precise knowledge of complex dielectric constant ($e_r = e' - je''$) as a function of frequency and temperature for the substrate material is necessary. The magnitude of the real and imaginary components of the complex dielectric constant as a function of frequency and temperature must be measured with high accuracy. Measured real and imaginary components of dielectric con-

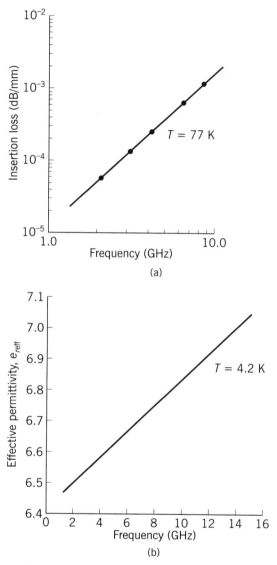

FIGURE 3.9 Insertion loss and permittivity values for various substrates and films as a function of frequency: (a) insertion loss for a YBCO film on MgO substrate at 77 K; (b) frequency dispersion of a superconducting niobium microstrip line on a sapphire substrate.

stants for the magnesium oxide and sapphire hard substrates as a function of frequency are shown in Figures 3.10 and 3.11, respectively, at 300 and 20 K [6]. In each case, thin films of YBCO superconducting material were involved. The real parts of the complex permittivity parameter at 300 K decrease as the

52 SUPERCONDUCTING SUBSTRATE MATERIALS

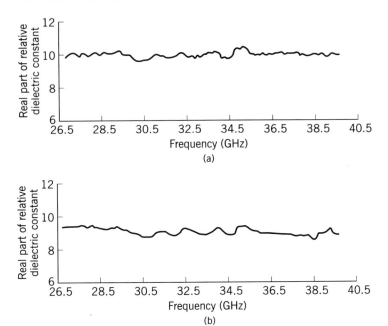

FIGURE 3.10 Measured values of real part of complex permittivity for the MgO substrate at (a) 300 K and (b) 20 K [6].

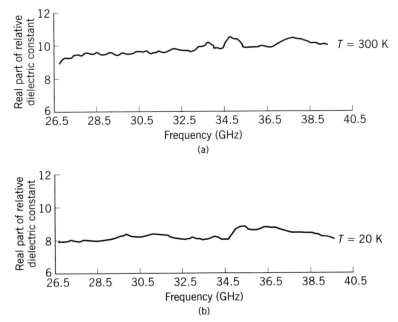

FIGURE 3.11 Real component of relative dielectric constant for the sapphire substrate as a function of temperature and frequency [6]: (a) room temperature; (b) 20 K.

3.3 HARD DIELECTRIC SUBSTRATES

operating temperature is reduced over the entire frequency range from 26.5 to 40.0 GHz.

The variations in the real component for both substrate materials are fairly smooth from 26.5 to 34.5 GHz, which represents about 26% of the band. In Figure 3.10, the variation in the value of real component of permittivity is within 4% over the same band for the YBCO films on MgO substrate at 20 K, while the variation in Figure 3.11 is less than 1% for the YBCO film on the same substrate at 20 K over the same frequency bandwidth. This indicates that sapphire offers minimum anisotropy and maximum dimensional stability at MM-wave frequencies under cryogenic operations.

Absolute values of complex permittivity for HTSC substrate materials, incuding MgO, Al_2O_3, $LaAlO_3$, ZrO_2, and SiO_2 at 300, 150, 70 and 20 K, are shown in Table 3.3B. These calculations are based on the measured values of the real and imaginary parts of the complex dielectric constants for various substrates identified in the upper section of Table 3.3A. Projected RF performance data as a function of frequency and temperature for a YBCO film on a MgO substrate are shown in Table 3.4.

It is evident from the measured and calculated data that the magnitude of both the real component and the absolute value of the complex permittivity parameter decreases for all the substrates investigated as the temperature is reduced. In case of MgO, the decrease in real component magnitude is about

TABLE 3.3A Real and Imaginary Components of the Dielectric Constants for Potential HTSC Substrates (measured at 33 GHz)

Temperature (K)	MgO e'_r	MgO e''_r	Al_2O_3 e'_r	Al_2O_3 e''_r	$LaAlO_3$ e'_r	$LaAlO_3$ e''_r	ZrO_2 e'_r	ZrO_2 e''_r	SiO_2 e'_r	SiO_2 e''_r
300	9.88	0.556	9.51	0.675	21.9	1.70	25.4	1.72	3.82	0.516
150	9.45	0.726	8.52	0.925	21.6	1.48	23.6	1.75	3.80	0.159
70	9.26	0.351	8.19	0.695	19.7	2.98	22.0	2.50	3.78	0.688
20	9.19	0.420	8.11	0.613	18.8	3.71	21.6	2.23	3.75	0.298

TABLE 3.3B Absolute Value of the Relative Dielectric Constant (e_r) e_r = sq. root ($e'^2_r + e''^2_r$)

Temperature (K)	MgO e'_r	Al_2O_3 e'_r	$LaAlO_3$ e'_r	ZrO_2 e'_r	SiO_2 e'_r
300	9.895	9.534	21.966	24.458	3.855
150	9.478	8.570	21.651	23.665	3.803
70	9.267	8.219	21.651	22.142	3.841
20	9.199	8.134	19.163	21.715	3.762

TABLE 3.4 Projected RF Performance Data for a High-Temperature Superconducting Ceramic Compound YBCA Film Deposited on MgO Substrate ($e_r = 10$)

		20 GHz		44 GHz		60 GHz	
T (K)	Resistivity (Ω-cm)	R_s (Ω)	L (dB/cm)	R_s (Ω)	L (dB/cm)	R_s (Ω)	L (dB/cm)
77	4.59×10^{-6}	0.060	1.06	0.089	1.57	0.104	1.84
65	1.63×10^{-6}	0.036	0.63	0.053	0.93	0.062	1.09
60	0.96×10^{-6}	0.028	0.52	0.042	0.77	0.048	0.90
40	0.90×10^{-6}	0.026	0.50	0.041	0.74	0.046	0.88

7%, while it is 15% for the Al_2O_3 at 34.5 GHz, when the temperature is reduced from 300 to 20 K. The decrease in the real component is 14% for $LaAlO_3$ and 15% for ZrO_2 material, respectively, under similar temperature conditions. The decrease in the absolute value is less than 1.8% at 34 GHz for the fused silica (SiO_2) substrate, when the temperature is reduced from 300 to 20 K.

3.3.3 Projection of Insertion Losses

The overall insertion loss in a microstrip transmission line or circuit is the sum of the conductor loss, substrate loss, and reflection loss. In actual practice, the conductor loss and the substrate loss are the major sources of insertion loss. Since the surface resistance of thin films of HTSC materials is on the order of a few milliohms at cryogenic temperatures, the conductor loss is much less than 5% at a temperature equal to or less than 50% of the critical temperature of the superconducting material. As a result, most microwave losses are confined to the substrate medium. As mentioned earlier, both the conductor loss and the substrate loss increase with temperature and frequency at moderate speed. Rapid reduction in insertion loss has been noticed as the operating temperature approaches 50% of the transition temperature (T_c) of the superconducting material.

The insertion loss in a microstrip line involving thin films of YBCO on MgO substrate has been computed as a function of temperature at MM-wave frequencies (Table 3.4). These computed values agree very closely with the measured data [7] shown in Figure 3.12. Note that the computed value of 1.06 dB/cm at 20 GHz at 77 K is in close agreement with the measured value of 1.1 dB/cm at 19.2 GHz and 77 K as designated by dotted line in Figure 3.12. If minimum insertion loss is the principal requirement, it will be necessary to cool down the superconducting components operating at higher MM-wave frequencies (60 GHz or higher) well below 50% of the transition temperature.

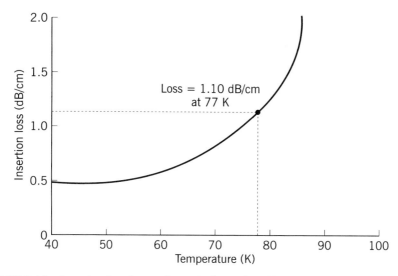

FIGURE 3.12 Insertion loss for a microstrip line using thin film of YBCO on MgO [7]. YBCO film thickness = 1 µm; MgO substrate thickness = 25 mils; frequency of measurement = 19.2 GHz; relative dielectric constant (e_r) = 10; temperature of measurement = 77K.

3.4 THE LATEST HARD SUBSTRATE MATERIALS FOR MICROWAVE CIRCUIT APPLICATIONS

In this section, the latest hard substrates for possible use in microwave and microelectronic circuits are identified. Rf properties of the latest monocrystal materials, such as $CaNdAlO_4$, $SrLaAlO_4$, $SrLaGa_3O_7$ and $NdGaO_3$, are briefly discussed. Some of these materials will be most suitable for deposition of superconducting thin films. The majority of these materials exhibit a uniaxial anisotropic property. Only strontium lanthanum gallate ($SrLaGa_3O_7$) offers room temperature (300 K) permittivity of 8.88 parallel to the optical c axis and 12.85 perpendicular to the c axis. This material has the lowest value of loss tangent over the temperature range of 0 to 100 K.

Permittivity decreases with temperature in all of the latest hard substrates, except $CaNdAlO_4$ (when the E field is perpendicular to the c axis). Microwave losses in $SrLaAlO_4$ and $SrLaGa_3O_7$ decrease with temperature, while the losses in $CaNdAlO_4$ and $NdGaO_3$ substrates increase at temperatures below 100 K, which makes them unsuitable for operations below 100 K. The magnetic interactions within the sublattice of the neodymium (Nd) ions with microwave field seem to be responsible for higher losses in $CaNdAlO_4$ and $NdGaO_3$. These two materials are not recommended for HTSC circuit applications operating at temperatures below 100 K.

Permittivity versus temperature and loss tangent versus temperature are plotted in Figures 3.13 and 3.14, respectively, for the $SrLaAlO_4$ and

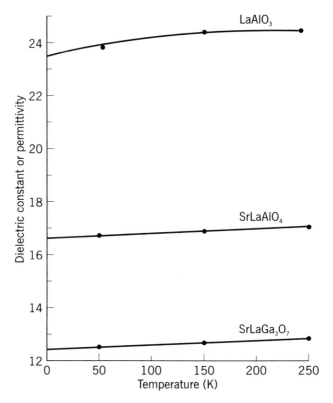

FIGURE 3.13 Measured dielectric constant for the latest HTSC substrate materials (SrLaGa$_3$O$_7$ and SrLaAlO$_4$) [8]. The curve for LaAlO$_3$ is shown for comparison.

SrLaGa$_3$O$_7$ materials. The dielectric constants of these two materials hardly experience any change from 0 to 100 K, thereby demonstrating the highest dimensional stabilities over this temperature range. Their properties are compared with those of the most popular hard substrate, LaAlO$_3$. The permittivity and loss tangent values of these two substrates shown in Figures 3.13 and 3.14 are valid only when the E field is perpendicular to the optical c axis [8].

Reference [8] summarizes the dielectric properties of LaAlO$_3$, SrLaAlO$_4$, and SrLaGa$_3$O$_7$ monocrystal based on experimental data collected over 4 to 40 GHz and a temperature range of 10 to 300 K. The data indicate that these three substrate materials are most suitable for precise designing of microwave superconducting components and devices. These materials do not contain magnetic ions, so the losses can be considered strictly as dielectric losses. The material SrLaGa$_3$O$_7$ exhibits the highest anisotropy and offers the lowest permittivity (when the E field is parallel to the c axis).

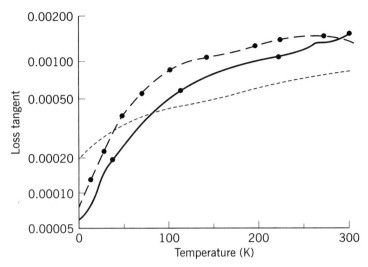

FIGURE 3.14 Measured values of loss tangents for the latest substrate materials as a function of temperature [8]: —, $SrLaGa_3O_7$; ---, $SrLaAlO_4$. $LaAlO_3$ (– – –) is shown for comparison. In all cases, the E-field is perpendicular to the optical c axis.

3.5 METALLIC SUBSTRATES FOR HIGH-POWER CIRCUITS

Recent research and development activities have stressed the need for special superconducting substrates capable of handling high power. Thallium-based superconducting thick films on metallic substrates appear to be the most suitable for meeting the high rf power requirements. Large-area Tl-Ba-Ca-Cu-O (TBCCO) thick films, when deposited on silver alloy substrates through a magnetron-sputtering process, have demonstrated potential applications in high-power microwave cavities. The best surface resistance values of such films achieved to this date are 4 and 14 mΩ at 10 and 77 K, respectively. Corresponding values of copper films are 8 and 21 mΩ at the same temperatures.

The surface resistance of TBCCO thick films begins to rise at 1–10 Oe of magnetic field and saturation generally occurs from 20 to 60 Oe. Thick films of TBCCO with a high degree of c axis texturing exhibit the weakest dependence of surface resistance on microwave power. These films also exhibit the sharpest transition to the superconducting state at high-frequency operations.

Absorption measurements at high microwave power by large-area thallium-based superconducting thick films on metallic substrates have demonstrated the weakest dependence of surface resistance (R_s) on the magnetic field produced by microwave power [9]. Absorption measurements were made using thick films of YBCO and Nb superconducting materials. These measurements indicated that the magnitude of R_s for polycrystalline thallium films under low magnetic field conditions was on the order of a magnitude higher than for YBCO films on MgO substrate and more than two orders of magnitude higher than for niobium (Nb) films. A bulk polycrystalline material exhibited higher values of R_s than those of polycrystalline-textured thick films. The measured R_s [9] values were higher than those for epitaxial thin films. In conclusion, thick films of thallium magnetron-sputtered onto a silver alloy (CONCIL 995) substrate are best suited for use in high-power microwave cavities.

SUMMARY

HTSC substrate materials can be classified in three major categories: soft, hard, and metallic substrates. Soft substrates, incuding ROHACELL-71, CuFlon, pure Teflon, microfiber PTFE, and woven PTFE, are most suitable for passive and low-power superconducting microwave circuits and components. Hard substrates, such as sapphire, alumina ceramic, quartz, fused silica, strontium titanate, magnesium oxide, zirconium oxide, and lanthanum aluminate, are best suited for applications where excellent dimensional stability, high-power handling capability, and high mechanical integrity are the principal requirements. The latest hard substrates such as strontium lanthanum aluminate ($SrLaAlO_4$) and strotium lanthanum gallate ($SrLaGa_3O_7$) can be used where minimum dielectric loss parallel to the c axis is the basic requirement. Metallic substrates are best suited for high-power microwave cavities using thick films of TBCCO material.

Recently discovered mercury-based HTSC compounds, namely, $HgBa_2Cu_4O_{4+x}$, $HgBa_2CaCu_2O_{6+x}$, and $HgBa_2Ca_2Cu_3O_8$, have demonstrated a transition temperature range from 93 to 135 K. However, their potential applications in microwave and microelectronic circuits have not been identified.

The selection of the right substrate strictly depends on the performance requirements, operating conditions and fabrication cost. Critical parameters, such as permittivity, loss tangent, substrate thickness, coefficient of contraction, anisotropy, and relevant mechanical strengths, must be considered for the selection of the right substrate for specific microwave circuits. A substrate with a low dielectric constant, lowest loss tangent, nearly unity anisotropy, high dimensional stability, matching coefficient of contraction, and fine grain structure must be given top consideration in the selection process.

REFERENCES

1. T. Laverghetta. Microwave materials: the choice is critical. *Microwave J.*, 162–173 (September 1993).
2. Roger Corp., Chandler, AZ. *Technical Report*, 16 January 1979, pp. 20–25.

3. Mullard Radio Astronomy Observatory, University of Cambridge, Cambridge, UK. *Technical Report*, September 1983.
4. D. Arthur. Measured data on tensile modulus and compressive modulus at cryogenic temperatures. *Technical Report*. Chandler, AZ: Roger Corp., March 1986, p. 4.
5. J. Bybokas and B. Hammond. High-temperature superconductors. *Microwave J.*, 127–138 (February 1990).
6. F. A. Miranda and W. L. Gordon. Measurements of complex permittivity of microwave substrates in 20- to 300-K temperature range from 26.5 to 40.0 GHz. *Cryogenic Engineering Conference*, Los Angeles, CA, 24 July 1989.
7. J. H. Takomoto, F. K. Oshita, H. R. Patterman, P. Kobrin, and E. Soverworth. Microstrip ring resonator technique for measuring microwave attenuation in high-T_c superconducting thin films. *IEEE Trans. MTT* 37:1650–1652 (1989).
8. J. Konopka and I. Wolff. Dielectric properties of substrates for deposition of high-T_c thin films up to 40 GHz. *IEEE Trans. MTT* 40:2418–1423 (1992).
9. D. W. Cooke, P. N. Arendt, E. R. Roy, and A. M. Portis. Absorption at high microwave power by large-area Tl-based superconducting films on metallic substrates. *IEEE Trans. MTT* 39:1539–1544 (1991).

CHAPTER FOUR

Application of Superconducting Technology to Passive Components

Superconducting technology has potential applications in both active and passive components. In this chapter discussion is limited to passive components and circuits, with emphasis on rf and digital components and circuits. We must understand the electromagnetic (EM) properties of superconducting films and transmission line configurations before we can apply low-temperature and high-temperature superconducting technologies. The classical two-fluid model is used to characterize EM properties. Mathematical expressions for the complex dielectric constant, complex surface impedance, and other propagation parameters for the superconducting transmission lines and media are outlined. Potential applications of low- and high-temperature superconducting films to coplanar waveguide (CPW) filters, microstrip filters, printed-circuit antennas, microwave cavity resonators, digital circuits, MM-wave E-plane filters, HTSC microwave delay lines using Nb and YBCO films, SAW devices, IR detectors, bolometers, and chirp filters are discussed. These devices do not require bias voltages or currents to operate and, hence, are entirely passive.

4.1 DERIVATION OF SURFACE IMPEDANCE

Expression for the complex surface impedance ($Z(s)$) of a superconducting film is derived by modifying the Maxwell equations based on the two-fluid model and London penetration equation. The electrons in a superconducting film consist of normal electrons and superconducting electrons, based on the two-fluid model. Thus, the total current density can be expressed as

4.1 DERIVATION OF SURFACE IMPEDANCE

$$\vec{J} = \vec{J_n} + \vec{J_s} \tag{4.1}$$

where J_n and J_s are current density vectors for normal and superconducting electrons, respectively, in a superconducting medium. The current density vector due to normal electrons can be written as

$$\vec{J_n} = \vec{\sigma_n} + \vec{E} \tag{4.2}$$

where $\vec{\sigma_n}$ is the electrical conductivity due to normal electrons and represents the real components of the complex conductivity parameter. $\vec{E}$ is the electric field vector in the film. The complex conductivity is defined as

$$\vec{\sigma} = \vec{\sigma_n} - j\vec{\sigma_s} \tag{4.3}$$

The superconducting electrons are best described by the London equation [1]:

$$\frac{d\vec{J_s}}{dt} = \frac{\vec{E}}{\mu\lambda^2} \tag{4.4}$$

where μ is the permeability, and λ is the London penetration depth. This expression represents the free acceleration of the electrons in an electric field vector $\vec{E}$ associated with superconducting material. Equations (4.2) and (4.4) can be incorporated in the first Maxwell equation, which can be written as

$$\text{Curl } \vec{H} = \vec{E} \tag{4.5}$$

where

$$Y = \sigma + jwe \tag{4.6}$$

The current density of superconducting electrons is written as

$$J_s = \frac{1}{wm\lambda} \tag{4.7}$$

For a normal conductor, the superconducting component σ_s of complex conductivity is zero. The second Maxwell equation is written as

$$\text{Curl } \vec{H} = z\vec{H} \tag{4.8}$$

In this case the magnetic field vector remains unchanged and the equation yields

$$z = jw\mu \tag{4.9}$$

The time-averaged complex power density P_s [2] can be expressed as

$$P_s = -\tfrac{1}{2}\text{Div}\,\vec{E} \times \vec{H} \tag{4.10a}$$
$$= P_d + 2j(W_k + W_m - W_e) \tag{4.10b}$$

where P_d is the dissipated power density corresponding to the normal (real part) component of complex conductivity, while the imaginary component is related to the kinetic energy density of the superconducting electrons. The symbols used in equations are defined as follows:

W_k = time-averaged kinetic energy density
W_m = time-averaged stored magnetic energy density
W_e = time-averaged electron energy density

The kinetic and magnetic inductance of an infinite film of arbitrary thickness can be evaluated from the above equations. Now we describe the electric and magnetic fields perpendicular to the film. The magnetic and electric field intensities inside superconducting film with a plane wave propagating in the z direction (perpendicular to the film surface) and with film boundaries located at $z = 0$ and $z = t$, can be written as

$$H_y(z) = H_{01}[\exp(-jkz)] + H_{02}[\exp(jkz)] \tag{4.11a}$$

$$E_x(z) = n\{H_{01}[\exp(-jkz)] - H_{02}[\exp(jkz)]\} \tag{4.11b}$$

where k is wave number ($\sqrt{-zy}$) and n is intrinsic impedance ($\sqrt{z/y}$). For a normal conductor, the wave number can be written as

$$k = \frac{\sqrt{w\mu\sigma}}{2}(1-j) = 1 - \frac{j}{\delta} \tag{4.12a}$$

$$n = \sqrt{\frac{w\mu}{2\sigma}}(1+j) = \frac{w\mu\delta}{2}(1+j) \tag{4.12b}$$

where δ is the classical skin depth, which is written as

$$\delta = \sqrt{\frac{2}{w\mu\sigma_n}} = \frac{1}{\sqrt{\pi\mu f \sigma_n}} \tag{4.13}$$

For a superconductor, assuming $\sigma_n \ll \sigma_s$, the above parameters can now be written as

4.1 DERIVATION OF SURFACE IMPEDANCE

$$k = \sqrt{w\mu\sigma_s}\left(\frac{\sigma_n}{2\sigma} - j\right) \quad (4.14a)$$

$$n = \sqrt{\frac{w\mu}{\sigma_s}}\left(\frac{\sigma_n}{2\sigma} + j\right) \quad (4.14b)$$

In a superconductor, the power losses are generally much smaller than the stored energy and also are much smaller than the losses in a normal conductor. If we invoke the tangential continuity of the magnetic field vector $\vec{H}$ at the boundary, the complex power applied per unit area of the film can be written as

$$P_f = \tfrac{1}{2}Z_{s_1}|\vec{H}_1 - \vec{H}_2|^2 + \tfrac{1}{2}Z_{s_2}\left(|\vec{H}_1|^2 + |\vec{H}_2|^2\right) \quad (4.15)$$

where Z_{s_1} and Z_{s_2} are the surface impedance parameters, which are defined as

$$Z_{s_1} = \frac{n}{j\sin(kt)} \quad (4.16a)$$

$$Z_{s_2} = n\frac{\cos(kt-1)}{j\sin(kt)} \quad (4.16b)$$

Two limiting cases are considered, one for a very thin film and the other for a relatively thick film. For a very thin film (where the thickness t is much less than the skin depth), surface impedance parameters are reduced to

$$Z_{s_1} = \frac{n}{jkt} = \frac{1}{\sigma t} \quad (4.17a)$$

$$Z_{s_2} = \tfrac{1}{2}jw\mu t \quad (4.17b)$$

For a thick film ($t \gg \delta$), these parameters are reduced to

$$Z_{s1} = 0 \quad (4.18a)$$

$$Z_{s_2} = n \quad (4.18b)$$

But the complex impedances are normally expressed as

$$Z_{s_1} = R_{s_1} + jX_{s_1} \quad (4.19a)$$

$$Z_{s_2} = R_{s_2} + jX_{s_2} \quad (4.19b)$$

where R_s and X_s represent surface resistance and reactance, and subscripts 1 and 2 represent two film boundaries.

With magnetic field contributions at both film boundaries and when $t \ll \delta$, one can simply write the intrinsic impedance n as

$$n = R_s + jX_s \tag{4.20}$$

Computed values of resistive and reactive components of parameters Z_{s_1} and Z_{s_2} normalized to the quantities of bulk material are plotted in Figure 4.1.

4.2 PROPAGATION CHARACTERISTICS OF SUPERCONDUCTING MICROSTRIP TRANSMISSION LINES

Since most of the microwave components and circuits employ microstrip transmission line structures, we must become familiar with the propagation characteristics of a microstrip transmission line structure shown in Figure 4.2. The propagation parameters are derived based on the two-fluid model of superconductivity and the phenomenological equivalence method (PEM). The PEM method can be applied to other microwave structures to evaluate the losses. This method includes both the electric field penetration and a geometrical factor into the formulation. The surface impedance [2] in most simplest form can be written as

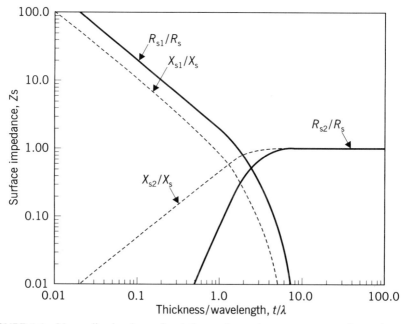

FIGURE 4.1 Normalized values of resistive and reactive components of complex surface impedances as a function of bulk material thickness [2].

4.2 PROPAGATION OF SUPERCONDUCTING MICROSTRIP TRANSMISSION LINES

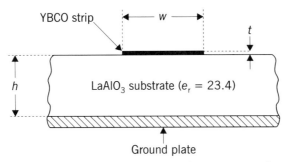

FIGURE 4.2 Cross-section view of a microstrip line structure using HTSC thin strip: W = HTSC strip linewidth, t = strip thickness, h = substrate thickness, e_r = relative permittivity.

$$Z_s = \frac{\sqrt{jw\mu}}{\sigma_s} \quad (4.21)$$

where σ_s is the complex conductivity of the superconductor, which can be defined in another format as

$$\sigma_s = \sigma_1 - j\sigma_2 \quad (4.22)$$

where
$$\sigma_1 = \sigma_n \left(\frac{T}{T_c}\right)^4 \quad (4.23a)$$

$$\sigma_2 = \frac{1 - (T/T_c)^4}{\lambda_0^2 w\mu} \quad (4.23b)$$

σ_n = normal conductivity near the critical temperature T_c
λ_0 = zero-temperature penetration depth
T_c = critical temperature where resistance is zero

The series impedance introduced per unit by a superconducting microstrip line is given as

$$Z_{se} = R_{se} + jX_{se} = Z_s G \coth[(\sigma_1 - j\sigma_2)Z_s AG] \quad (4.24)$$

where G is the geometric factor of the microstrip transmission line structure, Z_s is superconducting impedance, and A is the strip cross section = wt, where w is strip width, and t is strip thickness. The conductivity ratio of the real and imaginary components [2] can be expressed as

$$\sigma_r = \frac{\sigma_1}{\sigma_2} = \frac{\sigma_n T_r^4}{(1 - T_r^4)\lambda_0^2 w} \tag{4.25}$$

where T_r is a temperature ratio (T/T_c). A binomial expansion of Z_{se} can yield real and imaginary components and their ratio.

4.3 ATTENUATION CONSTANT (Ac) FOR A SUPERCONDUCTING LINE

The real part of the series impedance (Z_{se}) represents the attenuation constant (Ac), which is proportional to the square of the frequency. The imaginary part of the series impedance represents the propagation constant (A_p), which is a linear function of the frequency. However, both the components involve the geometric factor G of the structure and penetration depth (λ_0). For a simple transmission line equivalent circuit, the attenuation constant (Ac) can be written as

$$Ac = \frac{\text{Real part of } Z_{se}}{2Z_0} \tag{4.26}$$

Where Z_0 is the characteristic impedance of the lossless line. From the above equations, one gets the expression for this constant:

$$Ac = \frac{T_r^4}{(1 - T_r^4)^{1.5}} \frac{G\sigma_n}{4Z_0} \mu^2 \lambda_0^2 (\coth B + B \operatorname{cosec}^2 B) \tag{4.27}$$

where $B = AGX/\lambda_0$ and $x = 1 - T_r^4$. Computed values of the attenuation constant for a YBCO superconducting microstrip transmission line as a function of frequency are shown in Figure 4.3 based on various assumed parameters.

The surface impedance for a superconducting microstrip line is eloquently given by Reference [3], and can be written as

$$Z_s(f) = \frac{(1+j)\sqrt{(w\mu/2\sigma_n)}}{\sqrt{(\sigma_1 - j\sigma_2)/2}} \tag{4.28a}$$

and

$$Z_s(f) = R_s(f) + jX_s(f) \tag{4.28b}$$

where $R_s(f)$ and $X_s(f)$, the real and imaginary components, respectively, of the complex surface impedance, are function of operating frequency. Plotting the normalized values of the resistive and reactive components gives curves like those in Figure 4.1.

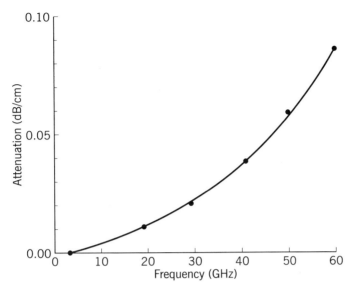

FIGURE 4.3 Computed values of attenuation as a function of frequency for a YBCO microstrip line on LaAlO$_3$ substrate [2]: $T = 77$ K, $T_c = 92.5$ K, $\sigma_n = 1.75$ µm, $\lambda_0 = 3000$ Å (London penetration depth).

The simulated cross section of an HTSC microstrip transmission line is illustrated in Figure 4.4. The current density distribution inside the strip as a function of width (along the x direction) and strip thickness (along the y axis) can be seen. The current density is based on the parameters assumed for a YBCO strip on LaAlO$_3$ substrate at cryogenic temperatures. Magnetic walls are employed to terminate the open boundaries. The current is carried by the superconducting surface adjacent to the dielectric substrate. The normalized current density decreases with the increase in y parameter, but it increases slightly as y approaches the strip thickness t. The quality factor, which is a function of film quality, its physical parameters, operating temperature, and critical temperature, decreases monotonously with the frequency, as illustrated in Figure 4.4.

4.4 dc CRITICAL CURRENT DENSITY IN THE MICROSTRIP LINE

The dc critical current density J (A/cm^2) is another performance parameter of the HTSC strip or film. The current density will vary from film to film for the same superconducting material, regardless of the strip thickness. The critical current density values of the BSCCO postannealed superconducting films are compared to those of YBCO postannealed films in Figure 4.5. It appears that

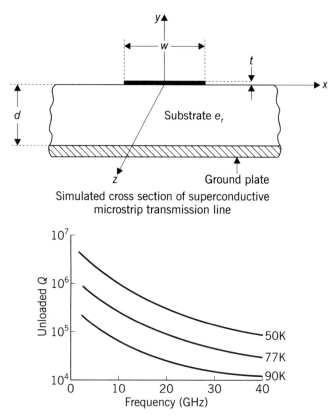

FIGURE 4.4 Unloaded Q and simulation of HTSC microstrip transmission line as a function of temperature and frequency: W = strip linewidth, d = substrate thickness. $W = 250\,\mu\text{m}$, $e_r = 24$, $d = 425\,\mu\text{m}$, $T_c = 100\,\text{K}$, $t = 1\,\mu\text{m}$, $\lambda_0 = 1800\,\text{Å}$.

the current density is enhanced at lower annealing temperatures. Critical current densities of the c-axis-oriented TBCCO thick film and polycrystalline YBCO thin films are also plotted in Figure 4.5 as a function of cryogenic temperature. These curves allow a microwave design engineer to select an optimum film material capable of meeting the desired performance level of the device.

4.5 PASSIVE MICROWAVE COMPONENTS USING HTSC TECHNOLOGY

Having derived the expressions for the attenuation constant, phase constant, current density, and dissipation factor, we can now proceed to the application of HTSC technology to passive microwave components, such as diplexing filters, bandpass filters, delay lines, digital circuits, infrared detectors, and

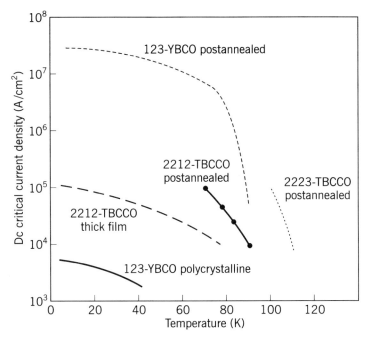

FIGURE 4.5 Theoretical current densities for various HTSC materials prepared using different processing techniques [4].

printed-circuit antennas. The application of HTSC technology to active microwave microwave components and devices is discussed in Chapter 5.

4.5.1 Frequency Multiplexing Filters

The initial multiplexing filter pairs were fabricated using low-temperature superconducting (LTSC) niobium (Nb) thin films on a sapphire substrate ($e_r = 9.4$). The use of sputtered Nb thin films for these filters was viewed as risk-free from the material standpoint and it was possible to obtain good-quality films on a large single crystal of an anisotropic sapphire dielectric substrate. A pair of parallel-coupled stripline resonator filters were designed and fabricated at 10 GHz. The films were deposited onto a rectangular sapphire substrate with dimensions not exceeding 2.54 cm × 1.27 cm × 0.38 cm and with the c axis parallel to the long dimension. The parallel-copupled resonators were obtained photolithographically parallel to the c axis. The electrical fields were mostly confined to the planes perpendicular to the c axis. The rf performance of the LTSC filter at 4.2 K was significantly improved over the room temperature performance level.

An example of a multiplexing filter involves an X-band filter pair with Tchebychev response (0.1 dB ripple) employing 6 poles and 7 microstrip-coupled quarter-wave sections. The passband responses and skirt selectivities of the multiplexing filters using YBCO films on $LaAlO_3$ substrate at 77 K are shown in Figure 4.6. The center frequency of the low-frequency filter is 9673 MHz, while the center frequency of the high-frequency filter is 9894 MHz. The insertion loss is less than 1.5 dB and the stopband rejection is greater than 60 dB for both filters [4].

4.5.2 Bandpass Filters

Several bandpass filters have been developed by various companies and tested at cryogenic temperatures using both normal and superconducting ground planes. Bandpass filters, all operating in X-band, have 4-pole parallel-coupled resonator microstrip configurations. The superconducting filter employs thallium thin films on $LaAlO_3$ substrate ($e_r = 24$, loss tangent = 0.00004, and $Q_{un} = 25,000$ at 77 K). In Figure 4.7, the performance of the superconducting filter is compared with performance levels of other filters at 77 K, using waveguide, combline, and microstrip structures. Microstrip filter structures possess

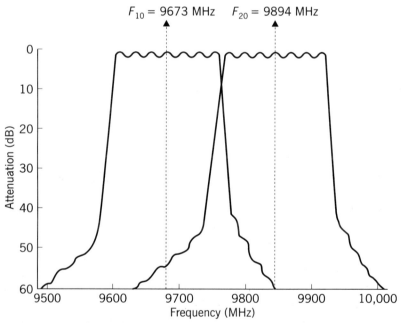

FIGURE 4.6 Frequency responses for two superconducting multiplexing filters at 77 K. The center frequency for the lower-frequency filter is 9673 MHz and that for the higher-frequency filter is 9894 MHz. The insertion loss is about 1.5 dB and stopband rejection is more than 60 dB at the required out-of-band frequencies for both bandpass filters.

4.5 PASSIVE MICROWAVE COMPONENTS USING HTSC TECHNOLOGY

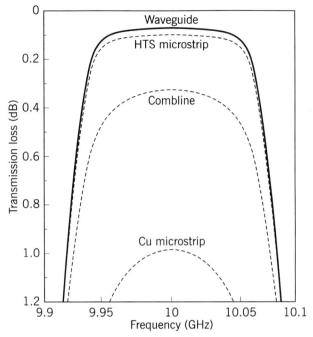

FIGURE 4.7 Performance comparison of the superconducting filter with other bandpass filters at 77 K. The HTSC microstrip filter uses thin films of thalluim on lanthanum aluminate substrate. All filter designs use 5 poles. [Courtesy of Superconductor Technologies Inc., Santa Barbara, CA.]

a unique advantage in terms of performance capabilities. Superconducting technology offers a substantial reduction in insertion loss, particularly, in narrow-band microwave and MM-wave filters. Good impedance matching networks using HTSC technology at the input and output ports of the microwave filters may be required if high return losses are desired in superconducting bandpass filters.

Microwave filters employing HTSC technology offer not only minimum insertion loss, but also considerable reduction in weight and size of the filter assembly. The insertion loss and the out-of-band rejection can be further improved by cooling the filter assembly down to 4.2 K. Microwave filters with low insertion losses are used in radars, satellite communication systems, ground receiving stations, and ECM equipment.

The measured response of a 3-pole, 10-GHz, microstrip bandpass filter using thallium films on lanthanum aluminate, designed and developed by Du Pont (Wilmington, DE), is illustrated in Figure 4.8. The bandpass filter demonstrates an insertion loss less than 0.3 dB and stop-band rejection greater than 25 dB at 77 K and at 10 GHz. Measured performance data on

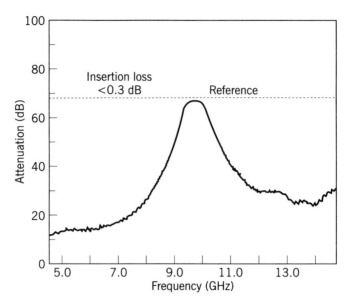

FIGURE 4.8 Response of a 3-pole, 10-GHz superconducting bandpass filter using thallium film on lanthanum aluminate. Insertion loss = 0.3 dB (max) at 77 K and 0.1 dB (max) at 4.2 K, passband VSWR = 1.50:1 (max), and stopband loss = 30 dB (min). [Courtesy of DuPont Corp., Wilmington, DE.]

cryogenically cooled bandpass filters using thallium and gold resonators at 77 K are illustrated in Figure 4.9.

4.5.3 HTSC Low-Pass Filter Using a Coplanar Waveguide (CPW)

A coplanar waveguide is the most convenient thin film structure for filter applications, because only one side of the substrate needs to be coated with HTSC material, which minimizes the cost and complexity of the assembly. This allows the other side to be thermally attached to the surface during the deposition process and in situ annealing of the superconductor. CPW resonators made from YBCO have demonstrated much lower losses than copper at 77 K. Measured data on low-pass filters using CPW structures exhibited lower insertion losses than those of filters using copper or silver at 77 K [5].

The attenuation for a CPW fabricated with YBCO on $LaAlO_3$ substrate at 77 K and 9 GHz was less than 0.2 dB/cm for a center conductor width of 5 μm, 0.06 dB/cm with a width of 45 μm and 0.05 dB/cm with a width of 196 μm. Thus, minimum insertion loss in HTSC microwave filters is possible through CPW structure with larger center conductor width. A well-designed low-pass filter with CPW structure will yield adequate suppression over a wider stopband region with minimum insertion loss at cryogenic temperatures. HTSC CPW filters using YBCO films on $LaAlO_3$ substrate are used for microwave

4.5 PASSIVE MICROWAVE COMPONENTS USING HTSC TECHNOLOGY

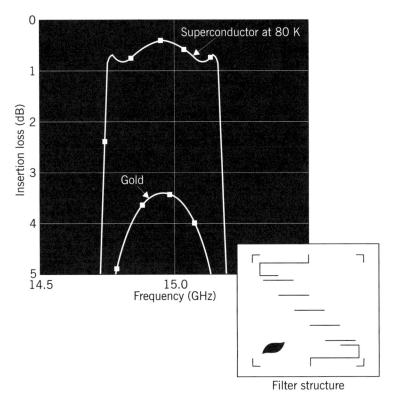

FIGURE 4.9 Filter performance comparison for superconducting thallium and gold films at 80 K. Insertion loss = 0.75 dB at 80 K and 0.25 dB at 4.2 K (estimated), and passband VSWR = 1.5 : 1 (max).

integrated circuits (MICs). The phenomenological equivalence method (PEM) model for superconducting films of thickness comparable to skin depth can be used with commercially available microwave CAD software to design CPW filters with minimum time and cost. The PEM model requires complete knowledge of London penetration depth, known as effective penetration depth, as a function of temperature. London penetration depth increases with temperature, as shown in Table 4.1.

4.5.4 Superconducting MM-Wave Filters

Limited performance data are available on MM-wave filters using HTSC microstrip resonators and CPW structures. The insertion loss, which includes the losses in the metallic strip and the substrate, increases with frequency, but the rate of increase is slow at 4.2 K operating temperature compared to 77 K. Estimates of insertion loss in HTSC MM-wave filters using microstrip structures as a function of temperature and frequency are summarized in Table 4.2.

TABLE 4.1 London Penetration Depth (λ_T) for YBCO and TBCCO HTSC Films as a Function of Temperature

Temperature (K)	YBCO ($\lambda_0 = 6800$)	TBCCO ($\lambda_0 = 4700$)
0	6,800	4700
4.2	6,801	4701
20	5,808	4704
40	6,830	4759
50	7,018	4848
60	7,551	5023
70	8,539	5359
80	10,716	6036

Note. The increase in penetration depth is about 3.13% for YBCO, compared to 3.15% for the TBCCO at T/T_c ratio of 0.5. Optimum performance of HTSC components is generally seen when operating temperature is less than 50% of critical temperature.

TABLE 4.2 Projected RF Performance Data for a High-Temperature Superconducting Ceramic Compound YBCA Film Deposited on MgO Substrate ($e_r = 10$)

T (K)	Resistivity (Ω-cm)	20 GHz $R_s(\Omega)$	20 GHz L (dB/cm)	44 GHz $R_s(\Omega)$	44 GHz L (dB/cm)	60 GHz $R_s(\Omega)$	60 GHz L (dB/cm)
77	4.59×10^{-6}	0.060	1.06	0.089	1.57	0.104	1.84
65	1.63×10^{-6}	0.036	0.63	0.053	0.93	0.062	1.09
60	0.96×10^{-6}	0.028	0.52	0.042	0.77	0.048	0.90
40	0.90×10^{-6}	0.026	0.50	0.041	0.74	0.046	0.88

Parameters assumed:
Z_o = 50Ω, $h = 0.004$ inch
q = dielectric filling factor (0.833)
e_r = relative dielectric constant (10)
e_{eff} = effective dielectric constant (8.5)
tan δ = loss tangent (0.005 assumed)
L_d = 0.26 db/cm (tan $\delta = 0.005$) at 20 GHz
    = 0.52 dB/cm (tan $\delta = 0.010$) at 20 GHz

The assumed parameters for the microstrip structure comprising YBCO resonators on MgO substrate are clearly specified. These losses are considerably lower for the microstrip lines made of YBCO films on LaAlO$_3$ substrate, because the loss tangent of this substrate is less than 0.00005.

Recent research and development activities on HTSC MM-wave filters indicate that the E-plane filter structure shown in Figure 4.10 offers the most promising filter design configuration. Experimental data on a 3-pole, E-plane bandpass filter incorporating a metal insert and fabricated from a high-T_c superconducting bulk material, YBCO, indicated an insertion loss less than 1.8 dB at 77 K and 35 GHz and stop-band rejection greater than 30 dB. This

4.5 PASSIVE MICROWAVE COMPONENTS USING HTSC TECHNOLOGY

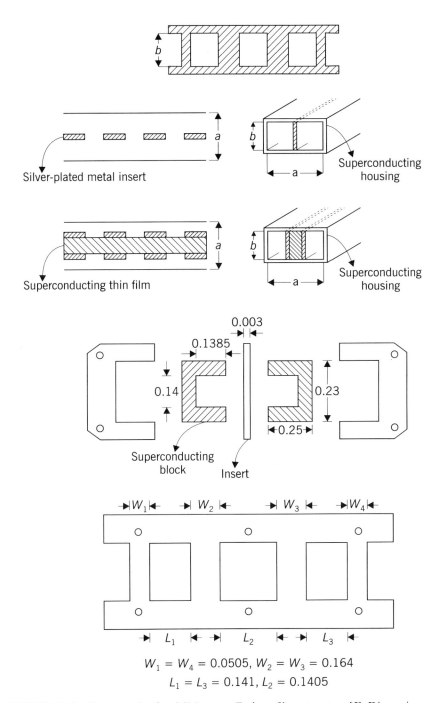

FIGURE 4.10 Superconducting MM-wave, E-plane filter structure [5]. Dimensions are in cm.

filter employed two U-shaped YBCO blocks inserted into a housing with a silver-plated metal insert. The rf performance of this E-plane filter can be further improved by using TBCCO thin film with reduced surface roughness on $LaAlO_3$ substrate and by operating at a much lower temperature. The insertion loss at MM-wave frequencies is largely dependent on surface roughness and uniformity of superconducting properties of the HTSC metallic surfaces. Quasi-optical [6] MM-wave bandpass filters using YBCO films on $LaAlO_3$ substrates are capable of operating over a 75 to 110-GHz frequency range. These filters offer a remarkable improvement in rf performance, particularly at 15 K.

4.6 SUPERCONDUCTING DELAY LINES

Conventional rf delay lines are made from waveguides, coax transmission lines, subminiaturized coax lines, microstrip transmission lines, coplanar waveguides, meander transmission lines, and exponential transmission lines. Conventional delay line structures suffer from excessive insertion loss of, weight, and size. However, substantial improvement in these parameters can be obtained by using high-quality superconducting films and low loss substrates. The author performed a brief tradeoff study in terms of weight, size, loss, pulse distortion, and frequency dispersion for coaxial conventional delay line structures and HTSC delay lines. Estimated values of weight, size, insertion loss, dispersion, and pulse distortion for a conventional coax delay line and HTSC delay line with a 100-ns delay are summarized in Table 4.3. An improvement of 100:1 in weight, size, and insertion loss for the HTSC delay line is evident. HTSC meander-line delay lines will offer further improvement in these areas because of their unique geometries. Calculated values of insertion loss for various delay lines as a function of temperature and frequency are plotted in Figure 4.11. The curve for the delay line fabricated from HTSC

TABLE 4.3 Performance Comparison for Various Delay Lines with a Superconducting Delay Line at 10 GHz

	Type		
Parameters	Coax Delay Line	HTST Delay Line	Estimated Improvement Factor
Delay (ns)	100	100	—
Loss (dB)	35	0.35	100
Weight (lb)	2.5	0.025	100
Size (cu. in.)	69	0.69	100
Frequency dispersion	Large	Negligible	—
Pulse distortion	Large	Minimum	—

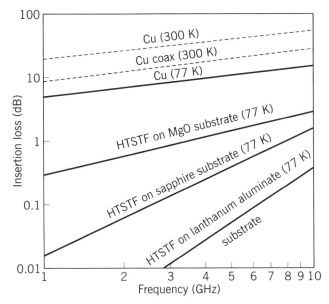

FIGURE 4.11 Insertion loss for various 100-ns delay lines as a function of temperature and frequency using high-temperature superconducting thin film (HTSCTF).

thin film on lanthanum aluminate substrate shows minimum insertion loss even at higher microwave frequencies and at 77 K. The ultralow insertion loss is due to the higher unloaded quality factors possible at cryogenic temperatures.

Projected values of unloaded Qs for an HTSC meander-line delay line as a function of temperature and frequency are shown in Figure 4.12. The highest unloaded Q is possible only at the lowest operating temperature, regardless of the operating frequency. HTSC delay lines have potential applications in airborne MTI radars and EW systems, because of minimum weight, size, and insertion loss.

The superconducting tapped-delay line has potential application in a superconductive chirp filter, which is widely used in complex radar signal processors. Passive superconducting filters, making use of delay lines called transversal filters, have potential application as matched filters in radar signal processors. Integration of superconducting technology and surface acoustic wave (SAW) technology in transversal filters offers matched filters with minimum insertion loss and size. Normal conductors such as copper and gold are unable to provide large delay with minimum loss, weight, and size. The superconductive transmission delay lines offer the lowest insertion loss and widest bandwidth because of their purely inductive nature.

A superconductive chirp filter with a dispersive time delay of 25 ns and a bandwidth of 3400 MHz centered at 4700 MHz has been fabricated using a niobium-on-silicon shielded microstrip technology. The filter assembly

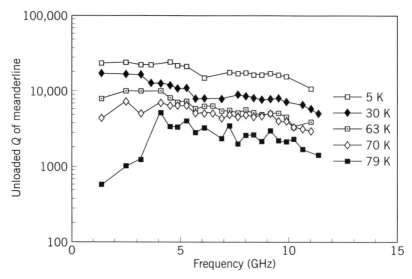

FIGURE 4.12 Unloaded quality factors of a superconducting meander line delay line as a function of temperature and frequency.

employed unequally spaced superconducting directional couplers as sampling devices along the delay line length. A superconductive shield mounted at a distance equal to the height of the dielectric on top of the microwave structure was used to achieve improvement in dispersion and coupler directivity over conventional microstrip structure. The conductive loss of this chirp filter fabricated with a 450-μm-thick niobium film was 0.012 dB/m and the dielectric loss was 0.014 db/m at 6.6 GHz and 4.2 K. This means that overall insertion loss of the chirp filter was 0.026 dB/m at 6.6 GHz and 4.2 K. This kind of rf performance is not possible with a conventional chirp filter. The estimated overall insertion loss for the superconducting filter was less than 0.104 dB at 26 GHz. This indicates that high-T_c superconducting technology can offer considerable performance improvement of the chirp filter operating at MM-wave frequencies. A comparison of SAW and high-T_c superconducting chirp filters is given in Table 4.3.

4.7 SUPERCONDUCTING RESONATORS

Superconducting microstrip ring resonators are used to measure superconducting material parameters such as surface resistance, film quality factor, and skin depth. Ring resonators are widely used because of their low radiation losses and minimum fabrication cost. Ring resonators offer precision techniques to measure the low-loss and dispersion properties of the HTSC films. The availability of low-loss epitaxial lanthanum aluminate substrate and the ability to deposit high-quality thin films on both sides of the same substrate permit the fabrication of microwave ring resonator circuits with true microstrip geometry. Superconducting ring resonators employing thin films of BaYtCuO (BYCO) and TaCaBaCuO (TCBCO) on lanthanum aluminate substrate were fabricated to measure the pertinent superconducting properties. Techniques using microstrip ring resonators and contactless probing schemes measure the critical parameters of superconducting materials with utmost accuracy and reliability. The critical parameters of the HTSC films have been measured with highest accuracy over a 2- to 22-GHz frequency range [7].

Ring resonators can also accurately measure the surface impedance of the microstrip circuit or transmission line. Both the resistive and reactive components of the superconducting surface impedance can be measured with high accuracy and repeatability. The resistive component contributes to the insertion loss in the superconducting circuit, whereas the reactive component represents the stored energy in the superconducting film. Microstrip as well as coplanar waveguide (CPW) YBCO resonators have been successfully employed to measure the surface resistance and penetration depth in the HTSC films with high accuracy and consistency. The surface reactance component slows down the electromagnetic (EM) wave. The stored kinetic energy of the charged carriers in superconducting film increases as the thickness of the film is reduced below the magnetic penetration depth. The influence of the surface reactance is used to determine the penetration depth in the HTSC films.

4.8 SUPERCONDUCTING SHIELDS

Superconducting enclosures are currently used as magnetic shields in magnetic resonance imaging (MRI) equipment to provide a high-quality image. There is a considerable demand for the superconducting shields operating at 77 K to screen the high-temperature superconducting quantum interference devices (SQUIDs) and other sensitive medical electronic equipment. In 1990, scientists at the University of Birmingham (UK) and at the AT&T Bell Labs (NJ) reported the behavior of the YBCO tubes for electromagnetic interference (EMI) screening capabilities for magnetic fields. Screening effectiveness is a function of superconducting tube dimensions, operating temperature, and frequency. Low-frequency screening is a serious problem for many applications. The shielding effectiveness of other HTSC mateials needs to be investigated as a function of temperature and frequency.

4.9 SIGNAL-PROCESSING DEVICES

Signal-processing functions are provided by a number of technologies, such as optical, digital, surface acoustic wave (SAW), integrated circuit (IC), and charge-coupled devices (CCD), which are readily available at reasonable costs. However, these devices suffer from bandwidth, processing time, and power consumption. Significant improvement in these areas is possible by using superconducting technology. Superconducting digital circuits offer high switching speeds, minimum gate delay (1.25 ps), low dissipation (12 µW), wide bandwidth, and ultralow signal dispersion. Josephson junction (JJ) logic gates, JJ 4-bit and 16-bit multipliers, and JJ 2-bit and 16-bit logic circuits when operated at cryogenic temperatures demonstrated minimum power consumption at maximum speed. Early in 1988, a 4-bit JJ microprocessor was developed capable of operating at 770 MHz. A year later, a 4-bit faster microprocessor was developed to operate an 1100-MHz clock with 100- µW power consumption. The higher clock rates with minimum power consumption that are possible only with low-temperature (4.2 K) superconductor technology are highly desirable for high speed A/D converters which can be used in advanced radars and electronic warfare systems.

4.10 APPLICATION OF HTSC TECHNOLOGY TO IR DETECTORS

Superconductivity technology can improve the performance of IR detectors, including photon detectors, bolometer detectors, photvoltaic detectors, and near-IR detectors, at cryogenic temperatures. The principal reason to cool a detector is to reduce the dark current level so that the signal-to-noise ratio (SNR) is no longer dominated by the dark current. Dark current is the electron current that flows in the detector material when the operating voltage is applied in the absence of optical radiation. Cooling of a detector is a complex issue, which depends on detector material properties, the band of operation, and the sensitivity requirement. Smaller versions of Stirling-cycle coolers, called stand-alone refrigerators, have been widely used to cool detectors below 200 K. Microcoolers are receiving great attention; because they weigh less than 300 g and consume less then 3.5 W of input power at 77 K. Microcoolers provide a net cooling capacity of 150 mW and are widely used to improve detector sensitivity.

4.10.1 Mercury Cadmium Telluride (Hg:Cd:Te) Detectors

Hg:Cd:Te detectors are the most popular and versatile IR detectors. They are capable of providing large dynamic range and optimum spectral response at wavelengths between 2 and 12 µm. These particular detector performance characteristics improve as the operating temperature is reduced below 77 K. Most photovoltaic Hg:Cd:Te cells operate over a 3- to 5-µm range, but at these wavelengths the devices have dark currents high enough to require Stirling-cycle cooling to maintain an operating temperature around 77 K for optimum detector performance.

Some Hg:Cd:Te detector devices incorporate dopants such as zinc that lowers the dark current level to allow operation between 180 and 230 K. IR detectors made of HgCdTe compound have demonstrated a detectivity greater than 10^{11} Jones (cm$\sqrt{Hz}$W^{-1}) at operating temperatures below 77 K, which is significantly higher than 10^{10} Jones at room temperature (300 K).

4.10.2 Indium Antimonide (InSb) Detectors

Indium antimonide (InSb) detectors offer reasonably good performance over 3- to 5-μm and 8- to 12-μm optical spectral ranges. InSb detectors, when operated below 77 K, demonstrated a background limited performance with a responsivity better than 10^{11} Jones. Even the 3-terminal, heterojunction, MM-wave HEMT devices, when operated at cryogenic temperatures, offer substantial improvement in noise figure and gain, particularly, at MM-wave frequencies.

4.10.3 Indium Gallium Arsenide (InGaAs) Detectors

InGaAs devices yield considerable performance enhancement at cryogenic temperatures. Starring focal-plane array devices made of photovoltaic InGaAs detectors demonstrated optimum performance over the 0.9- to 1.7-μm range when cooled below 77 K. These detectors offer a satisfactory performance even in daylight at room temperature, when they are used as thermal imaging sensors. For reduced light level operations, such as night vision imaging, a thermoelectric cooler is adequate. When the detectors are cooled down to 200 K, the SNR improves by a factor of 30:1 and detectivity exceeds 10^{14} Jones. Below 200 K, the SNR in the InGaAs detectors is background limited over the 0.9- to 1.7-μm spectral region and, thus, will provide no improvement in the detector performance level. Photovoltaic InGaAs detectors, when used as solar cells operating at 2.5 μm, must be cooled below 200 K to achieve detectivity better than 10^{14} Jones.

4.10.4 Gallium Arsenide (GaAs) and Aluminum Gallium Arsenide (AlGaAs) Detectors

At room temperature and above, a photoconductive positive-intrinsic-negative (PIN) detector made from GaAs semiconductor compound is sensitive to radiation over the 0.4- to 1.1-μm spectral range. The dark current in PIN detectors normally doubles for every 10-K increase in operating temperature. Low-temperature air cooling is sufficient to operate these devices with detectivity values close to 10^{14} Jones, except in high-duty applications involving thermal energy from short-wavelength infrared (SWIR) sources.

Detector devices with GaAs and AlGaAs quantum-well (QW) configurations operate at longer IR wavelengths (3–20 μm). Published experimental data reveal a detectivity value better than 10^{13} Jones at 9.2 μm when cooled to 35 K. In summary, IR detectors must be cooled well below 200 K, if optimum detector sensitivity is the principal requirement.

4.11 SUPERCONDUCTING RADIO-FREQUENCY ANTENNAS

The antenna is the most critical element of a radar or communication system or electronic warfare (EW) equipment. High radiation efficiency and low microwave circuit loss are of paramount importance regardless of the system operation. Electrically short antennas are preferred over quarter-wave or half-wave dipole radiating elements. Antenna efficiency depends on ohmic losses in the radiating elements, matching networks, and feed circuits. In electrically short antennas, the ohmic losses are significantly higher than other losses, including the radiation losses. The ohmic losses, which are dependent on antenna element resistance and matching network resistance, can be reduced by cooling the entire antenna structure down to liquid nitrogen temperature (77 K) or liquid helium temperature (4.2 K).

A significant increase in radiation efficiency has been realized by cooling the copper superconducting radiating elements down to 77 K; further improvement in radiation efficiency is possible if the elements are cooled down to 4.2 K. The use of rare-earth copper oxide superconducting materials in fabrication of rf antennas with critical temperatures well above the liquid nitrogen temperature have demonstrated significant improvement in the radiation efficiency and gain of the electrically short, printed circuit antennas, which are discussed in the next section.

4.11.1 High-T_c Superconducting Electrically Short Dipole Antennas

Antennas operating at VHF and UHF frequencies suffer from large physical dimensions and high ohmic losses. High-T_c superconducting short dipole antennas solve both problems. As mentioned before, the ohmic losses can be considerably reduced by cooling the antenna elements to 77 K or lower. An electrically short dipole antenna using a high-T_c superconducting, extruded ceramic YBCO compound wire, and an impedance matching network on a high-performance substrate was fabricated to operate at around 550 MHz. A similar antenna was fabricated using copper elements only with identical dimensions. Performance data measured on electrically short antennas indicate that the antenna made of superconducting material demonstrated higher radiation efficiency and gain over the same operating frequency band and at the same operating temperature. The radiation efficiency is dependent on the input impedance of the antenna and the radiation resistance of the elements. The input impedance of the antenna is the sum of ohmic resistance (R_o) and radiation resistance (R_r). Assuming a TEM wave propagation in the antenna, the rf current at the input to the dipole antenna can be calculated using appropriate equations, initially neglecting the line and dielectric support losses. The ratio of the input current to the dipole at 77 K to that at room temperature (300 K) can be obtained using appropriate expressions. The current ratio at 77 to 300 K predicts the gain improvement achieved using superconducting radiating elements.

The current ratio can be used to compare the gain improvement achieved using the equivalent copper antennas operating at 77 and 300 K. In Figure 4.13, a gain enhancement of 12 to 15 dB for the superconducting electrically short antenna using YBCO radiating elements is evident over the antenna using the copper radiating elements. The gain improvement for the YBCO antenna is about 6 dB over the identical copper antenna at 77 K. Additional gain enhancement is possible for the superconducting antenna if the operating temperature is further reduced to 4.2 K.

4.11.2 Superconducting Loop and Helical Printed-Circuit Antennas

For high-T_c superconducting printed-circuit antennas the theoretical radiation efficiency of both the copper loop and copper closed helical antennas is very small at room temperature. A loop antenna using YBCO wire along with

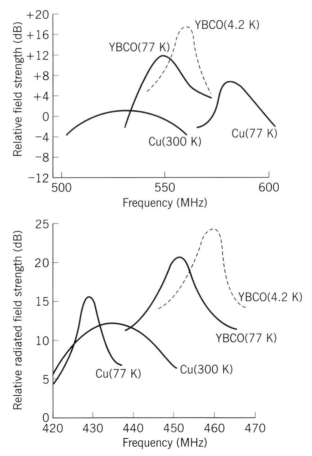

FIGURE 4.13 Performance levels of cryogenically cooled, electrically small antennas [8].

matching network was fabricated to operate at 450 MHz. The performance improvement of the loop antenna is expected to be similar to that of other antennas shown in Figure 4.13.

An electrically small, closed, helical antenna with few turns was also fabricated using YBCO wire. The total wire length of the helix was much less than one wavelength at the operating frequency. Structural details of this antenna are illustrated in Figure 4.14, but no performance data are available. Two copper antennas with the same dimensions were also fabricated to operate at 450 MHz over a temperature range of 77 to 300 K. The radiated power of the superconducting loop antenna was 9 dB higher at 77 K than that of the room-temperature copper loop, and 5 dB higher than that of the superconducting copper antenna at 77 K. The radiation efficiency of the supercooled copper loop antenna is roughly 32% and that of the room temperature copper loop is about 12.5% compared to 100% radiation efficiency of the YBCO dipole loop antenna shown in Figure 4.15. Performance improvement of half-loop antennas operating at cryogenic temperatures and using superconducting films of copper, YBCO, and TBCCO have been observed. The radiation efficiency of the 100-MHz half-loop antenna fabricated with YBCO and TBCCO films on lanthanum aluminate substrates is 12 to 20% higher than that of the copper antenna at the same temperature. The radiation efficiency of the copper loop is hardly 8%, compared to 18% for the YBCO loop and 38% for the TBCCO loop antenna at the same operating temperature of 20 K. The radiation efficiency of the TBCCO loop antenna may exceed 50% if the temperature is reduced to 4.2 K.

A superconducting integrated log periodic array antenna comprising of superconducting printed-circuit dipole elements has been fabricated. This antenna demonstrated higher gain, improved radiation efficiency, and symmetrical E-plane and H-plane beamwidths in addition to considerable reduction in weight and size, when operated at 77 K.

A cavity-backed printed-circuit spiral antenna was fabricated using 10-to 20-μm-thick films of YBCO on a high-performance substrate. This antenna

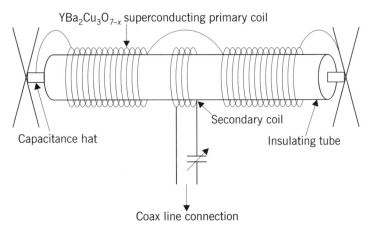

FIGURE 4.14 Cryogenically cooled, helical coil antenna using a YBCO wire element.

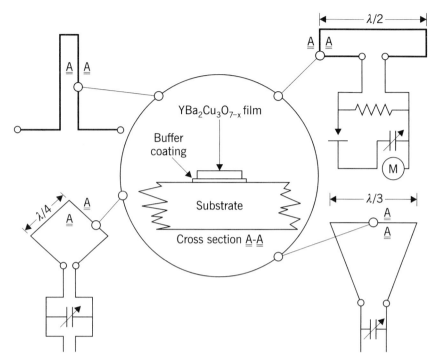

FIGURE 4.15 Cryogenically cooled, loop dipole antennas using superconducting thin and thick films.

demonstrated remarkable improvement in efficiency, gain, and pattern symmetry when operated below 77 K. No experimental data are avilable to support these claims, but test data on other superconducting antennas can fully justify the performance capabilities of the superconducting cavity-backed spiral antenna.

4.11.3 MM-Wave Superconducting Antennas

The superconducting antennas described so far were operated at low radio frequencies. Extensive experimental test data are not available on MM-wave superconducting antennas, but superconducting technology is expected to be very attractive in the design of MM-wave printed circuit antennas. Preliminary calculations indicate that a 35-GHz, 100-element linear dipole array with half-wavelength spacing can benefit significantly from an HTSC microstrip feed network and HTSC dipole radiating elements. The biggest payoff may occur at frequencies above 100 GHz, if the HTSC microstrip line with very thin dielectric spacings less than 100 μm thick can be fabricated and operated at 4.2 K. The efficiency–bandwidth product is fairly constant for an HTSC MM-wave antenna.

SUMMARY

Substantial reduction of insertion loss, weight, and size in superconducting passive rf components has been observed when devices are operated well below the critical temperatures of the HTSC materials used. Significant performance improvement in rf delay lines, bandpass filters, diplexers, and band reject filters has been noticed. Appreciable improvement in gain and radiation efficiency has been realized, particularly in superconducting electrically short rf antennas operating at lower frequencies, when cooled down to well below 77 K. Reduction in bandwidth has been observed in superconducting antennas because of higher unloaded Qs of the radiating elements at cryogenic temperatures. Substantial reduction in weight, size, and power consumption is possible for the digital circuits fabricated with HTSC materials. Detectivity and responsivity of IR detectors can be significantly improved by cooling the detector elements well below 77 K. Improvement in reliability due to implementation of superconductivity technology is possible in detectors and sensitive bolometers. Passive digital and optical components seem to realize considerable overall gain due to implementation of HTSC technology.

Radio-frequency systems or subsystems operating below 3 GHz are the prime candidates for the electrically short antennas [8] employing HTSC technology. In transmit antennas, power levels on the order of several kilowatts can be handled by the best existing HTSC materials. For receiving antennas, the performance improvement in terms of radiation efficiency and gain is possible only at operations above 30 MHz due to terrestrial noise level. The use of radiating elements and impedance matching networks made with HTSC materials can substantially improve antenna performance. Integration of HTSC technology has a significant impact on the resistive losses in the antenna elements, matching networks, and feed circuits. However, the dielectric losses due to loss tangent and permittivity parameters of substrates are not predictable. The loss tangent of the substrate must be kept below 0.0001 to minimize the substrate losses. An increase in radiation efficiency from 7 to 37% has been observed for a half-loop antenna using TBCCO thin films compared to an equivalent copper antenna, when operated well below 77 K. Superconductor Technologies Inc [9] designed and developed HTSC 10-GHz band pass and 4.6 GHz band-reject filters using thin films of TlCaBaCuO on $LaAlO_3$ substrate. These filters demonstrated minimum insertion loss with improved selectivity at 77 K operation.

REFERENCES

1. K. K. Mei et al. Electromagnetics of superconductors. *IEEE Trans. MTT* 39:1545–1552.
2. O. R. Baiocchi et al. Effects on superconducting losses in pulse propagation on microstrip lines. *IEEE Microwave Guided Wave Lett.* 1:2–4 (1991).

3. E. B. Ekholm et al. Attenuation and dispersion for high-T_c superconducting microstrip lines. *IEEE Trans. MTT* 38:387–394 (1990).
4. S. H. Talisa et al. Low- and high-temperature superconducting microwave filters. *IEEE Trans. MTT* 39:1448-1454 (1991).
5. W. Chew et al. Design and performance of a high T_c superconductor coplanar waveguide filter. *IEEE Trans. MTT* 39:1455–1460 (1991).
6. D. Zhang et al. Quasi-optical MM-wave bandpass filters using high-T_c superconductors. *IEEE Trans. MTT* 39:1493–1496 (1991).
7. C. Wilker et al. 5-GHz high-temperature superconductor resonators with high Q and low-power dependence up to 90 K. *IEEE Trans. MTT* 39:1462–1467 (1991).
8. R. J. Dinger et al. A survey of possible passive antenna applications of high-T_c superconductors. *IEEE Trans. MTT* 39:1498–1507 (1991).
9. M. S. Schmidt, et al. Measured performance at 77 K of superconducting microstrip resonators and filters, *IEEE Trans. MTT*, 39(9):1475–1479 (1991).

CHAPTER FIVE

Applications of Superconducting Thin Films to Active Rf Components and Circuits

Superconducting technology is useful when high sensitivity, low loss, low power consumption, high speed, small size, and low weight are the principal requirements. The Josephson junction (JJ) and superconductive quantum interference device (SQUID) are the building blocks and can play key roles in the design and development of passive and active components for analog, digital, microwave, and MM-wave applications. This chapter focuses primarily on the integration of HTSC and LTSC technologies into active components for possible applications in electromagnetic sensors, microwave/MM-wave subsystems, and satellite communication systems.

Active components and devices based on high- and low-temperature superconducting technologies may be used in conventional radar systems, phased-array radars, space sensors, electronic warfare equipment (EW), satellite communication systems, and optical systems. The most critical components and devices include superconducting phase shifters known as distributed Josephson inductance (DJI) phase shifters, HTSC/ferroelectric phase shifters, magnetically and electronically tunable filters such as YIG-tuned filters, superconducting rf switches, cryogenically cooled windows, and ferrite devices. Ferrite characteristics at cryogenic temperatures are described in this chapter. Critical design aspects and performance improvement of the components and devices due to integration of HTSC/LTSC technologies are identified.

5.1 PHASE SHIFTERS USING SUPERCONDUCTING TECHNOLOGY

A phase shifter is the most critical element of the phased-array antenna system. A superconducting phase shifter requires a special fabrication technique compared to that of a conventional phase shifter using either a diode or ferrite phase shifting element. The penetration of EM-waves into a superconducting surface differs from normal metals. It results in the kinetic induction effect, which offers high-power capability at an operating temperature very close to the critical temperature (T_c) of the superconductor material. In a very thin film significant slowing of the EM-wave propagation velocity occurs due to the change in the penetration depth, which offers true time-delay phase-shifter capability.

Low surface resistance, magnetic flux exclusion, and high critical current density (J_c) are the principal requirements for the superconducting phase shifters for application in high-power, electronically steerable, phased-array antenna systems. Optimum performance of any superconducting phase shifter is possible as long as the maximum operating temperature does not exceed 60% of the critical temperature of the superconducting film material.

Significant slowing of the EM-wave propagation velocity due to a change in penetration depth within a superconducting thin film is the basis of a superconducting phase shifter. The device is known as DJI phase shifter. Phase shifters using superconducting technology offer minimum insertion loss, small size and weight, low power consumption, true-time delay over extended bandwidth, and fast response time. Two phase shifters using superconducting technology are described in terms of design, performance, and limitations.

5.2 DISTRIBUTED JOSEPHSON INDUCTANCE (DJI) PHASE SHIFTER

The development of the DJI phase shifter [1] has been accelerated by aggressive research and development activities and the availability of HTSC device technology and circuits. The feasibility of the DJI phase shifter design was originally demonstrated using a low-temperature superconductor (LTSC) material called niobium (Nb) at 4.2 K. However, this particular Nb LTSC-DJI phase shifter required a large distributed array of phase-shifting elements. Development of superconducting Josephson junction devices and SQUIDs integrated thousands of such devices into a microstrip transmission line, which led to the development of a true time delay phase shifter. JJs and SQUIDs possess nonlinear inductance properties, which are most desirable for active microwave phase shifters. The nonlinear inductance (L_n) associated with a SQUID can be written as

$$L_n = L_{jj} L_s P_s \cos(2\pi\Phi/\Phi_0) \qquad (5.1)$$

where L_{jj} = Josephson junction inductance
         L_s = SQUID inductance P_s = SQUID parameter (≈ 1)
         Φ = magnetic flux in the SQUID
         Φ_0 = magnetic flux quantum, (2×10^{-15})V – s

Each SQUID device shown in Figure 5.1 [1] represents a variable inductor and acts as a lumped element transmission line. Thus, an inductively coupled SQUID represents a variable inductance, which provides both the phase shift and time delay in an rf or microwave circuit. The superconducting DJI phase shifter comprises a superconducting microstrip transmission line segment which is coupled to a large number of SQUIDs distributed along its length (Figure 5.1).

When integrated into a microstrip transmission line these SQUID devices change the velocity of the EM-wave by an amount equal to $1/\sqrt{Lc}$, where L and C are the inductance and capacitance (Figure 5.1). The magnetic flux in each SQUID device can be varied either by a dc magnetic field normal to the transmission line or by a current flowing into the microstrip line acting like a resonator. In a resonator, which contains hundreds SQUIDs, large phase can be achieved over a narrow rf bandwidth. Broadband true time-delay phase shifters are possible by EM coupling the superconducting microstrip transmission line to a large array consisting of thousands of SQUID devices.

5.2.1 Design Aspects of the HTSC DJI Phase Shifter

This phase-shifter design involves a monolithic circuit comprising several HTSC SQUID devices imbedded in a microstrip resonator (Figure 5.1). The JJs can be fabricated as a 90° grain boundary in an ABC superconducting thin film with sharp steps in the superconductive substrate lanthanum acuminate (LaAlO$_3$). The SQUID device comprises 2-µm-wide step-edge junctions and a pair of inductive square holes. The SQUID devices are duplicated and arranged in parallel across the microstrip transmission linewidth (W). Several columns of SQUIDs are laid out along the effective length of the microstrip transmission line with appropriate dimensions.

5.2.2 SQUID Geometry and Dimensional Parameters

Specific details on the SQUID geometry, various dimensions of the step edge, square holes with side dimension of a, hole-to-hole separation of d, substrate height of h above the ground plane, and superconductive thin-film thickness of t are shown in Figure 5.1. The square-hole pairs are joined by a 2-µm superconducting microbridge line that steps over a 0.25-µm LaAlO$_3$ substrate step. Based on the above-mentioned SQUID geometry and a microstrip transmis-

5.2 DISTRIBUTED JOSEPHSON INDUCTANCE (DJI) PHASE SHIFTER

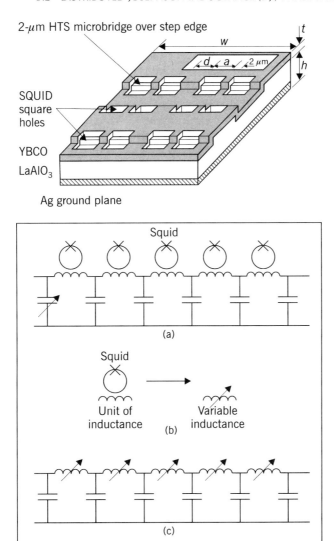

FIGURE 5.1 Design of a superconducting microwave phase shifter [1]. I_0 = current flux per quantum, a = side of the square hole, d = separation of holes, w = width of the microstrip, h = substrate thickness. For low coupling limit, $I_0 = 2\pi w \Phi_0 d / 3\mu_0 a^3$, where μ_0 is the permeability of the ferrite material, and Φ_0 is the quantum flow.

sion line width w, the expression for nonlinear inductance can be rewritten as follows:

$$L_n = L_0\left(1 + F_i \cos\frac{2\pi\Phi}{\Phi_0}\right) \quad (5.2)$$

where L_0 = sum of microstrip line inductance and JJ inductance coupled to the microstrip transmission line
F_i = interaction factor not exceeding 0.10

The interaction factor for a microstrip transmission line geometry can be easily calculated using the following equation:

$$F_i = \frac{7.5a^2 P_s}{h(a+d)} \qquad (5.3)$$

where a = side dimension of the square hole
h = substrate thickness
d = hole-to-hole spacing or separation
P_s = SQUID parameter not exceeding unity

Calculated values of the interaction factor as a function of parameters a, d, and h are shown in Table 5.1, assuming unity for the SQUID parameter.

5.2.3 Operating Principle

The operating principle of an electrically steerable phased-array antenna states that the relative phase shift between the two adjacent array elements must satisfy the following expression:

$$\phi = 2\pi \frac{s}{\lambda} \sin \theta \qquad (5.4)$$

where ϕ = reactive phase shift
λ = operating wavelength
s = spacing between the two adjacent elements

In a series-fed phased-array antenna, the two adjacent antenna radiating elements are connected by a superconducting phase shifter with phase shift ϕ. If

TABLE 5.1 Calculated Values of Interaction Factor F_i as a Function of Various Parameters

Substrate Thickness h (mils/μm)	Case A $a = 4\,\mu m$ $d = 6\,\mu m$	Case B $a = 5\,\mu m$ $d = 7\,\mu m$	Case C $a = 6\,\mu m$ $d = 8\,\mu m$
5/127	0.094	0.123	0.152
7.5/190	0.063	0.082	0.102
10/254	0.047	0.061	0.076
15/381	0.031	0.041	0.051
20/508	0.024	0.031	0.038

all the phase shifters are identical and introduce the same amount of phase shift, it is possible to achieve a total phase shift up to 360°. As stated earlier, phase shift in a DJI phaser is achieved by the Josephson inductance produced by the microstrip transmission line. The antenna beam is scanned electronically in the space as a function of phase shift produced by the DJI elements.

5.2.4 Conditions for Optimum Performance

It is evident from Equation 5.4 that electronic beam steering over wide scan angles requires a large number of DJI phase shifters. Maximum phase shift requires minimum substrate thickness h and large square-hole size compatible with power handling capability. Large square-hole size will require large hole-to-hole spacing d, leading to a large array. The optimum square-hole size is limited by two factors: the maximum rf power handling capability and the condition that the SQUID parameter (P_s) will not exceed 0.10, as shown in Table 5.1.

5.2.5 Performance Limitations

The DJI phase shifter comprises hundreds or thousands of SQUID devices integrated into a superconducting microstrip transmission line (SMTL), which can be controlled by external dc magnetic field. The amount of phase shift is dependent on the length of the SMTL, the amount of phase shift per SQUID element, which is limited by various factors, square-hole dimension, and the number of SQUIDs integrated into the SMTL. The useful range of external magnetic field is limited by the rf input power and by the periodicity of the SQUID phenomenon determined by the flux quantum parameter ϕ_0.

The impact of temperature on insertion loss and phase shift is shown in Figure 5.2. A constant peak-to-peak phase shift is seen within ±2° at input power of −30 dBm over the temperature range of 30 to 50 K. The phase-shift amount is limited by the saturation due to input power, which occurs at −10 dBm for this phase shifter. Phase variation less than a degree can be achieved with precision control of bias voltage and operating temperature. The 10-GHz DJI phase shifter shows a fairly constant insertion loss, less than 3 dB over the 4.2 to 85-K range (Figure 5.2a). The insertion loss of the same phase shifter can be reduced to well below 2 dB, if the device is operated below the critical temperature T_c of the superconductor used and with input power less than −20 dBm. The current per flux quantum is less than 400 µA and the peak-to-peak phase shift is less than 20°.

The amount of phase shift is limited by the microstrip line quality factor, which is around 6000 at 10 GHz and 77 K. Phase shift as large as 200° can be achieved in a high-Q resonant circuit. The phase shifter described here is most suitable for integration into the large phase-array antennas used in the space program. The power dependence of a high-density SQUID resonant phase

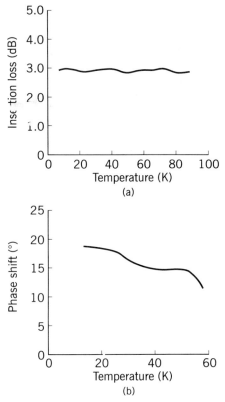

FIGURE 5.2 Impact of temperature on the performance of a 10-GHz superconducting phase shifter [2]: (a) insertion loss at input power of -20 dBm; (b) phase shift as a function of cryogenic temperature. The constant phase shift can be seen within $\pm 2°$ over 30–50 K at an input power level of -30 dBm.

shifter will be quite different than that of a low-density resonant phase shifter. This device is capable of providing optimum performance at 20 K and at input power not exceeding -10 dBm over an 8- to 18-GHz frequency range.

5.2.6 Performance Capabilities of Phased Arrays Using Superconducting Phase Shifters

A phased-array antenna using superconducting DJI phase shifters provides precision control of electronic beam steering and antenna sidelobe levels. Superconducting technology when integrated in beam-forming networks will reduce antenna losses and bias current requirements, thereby improving both the efficiency and reliability of the phased-array radar system. Due to an improved time constant at cryogenic temperatures, the electronic beam can be steered in much less than 1 µs. This ultrafast response of the superconducting DJI phase shifter can be important to a phased-array radar system designed to track thousands of targets under high-density target environments. Future phased-array radars will benefit from the outstanding capabilities of DJI phase shifters in terms of high resolution, fast response time, low power consumption, low insertion loss, and minimum weight and size. The integration of the DJI phase shifter in future space sensors, radar systems, satellite communication systems, and EW equipment will be aided by the availability of high-performance HTSC materials and low-cost cryogenic coolers. Future applications of superconducting DJI phase shifters include MM-wave beam-steering phased-array antennas, beam-forming networks for implementation in high-performance tracking radar systems, satellite communication systems, and sophisticated EW equipment.

5.3 HTSC/FERROELECTRIC PHASE SHIFTERS

This phase shifter is a hybrid design that employs two distinct technologies: HTSC and ferroelectric. In a hybrid HTSC ferroelectric (HHTSCF) phase shifter [1], a dc bias is required to change the dielectric constant or permittivity of the substrate and the capacitance per unit length of the HTSC microstrip transmission line. The ferroelectric property provides the variable lumped capacitance, as illustrated in Figure 5.3. An HHTSCF phase shifter is a true time-delay device, which is highly desirable in a coherent pulsed doppler radar using electronic beam steering for fast, multiple-target tracking.

5.3.1 Critical Design Aspects

In an HHTSCF phase shifter, the DJI transmission line [1] provides the variable inductance, and the ferroelectric material provides the variable capacitance (Figure 5.3). Sputtered or laser-ablated thin-film ferroelectric materials are used for fabrication on this component. The microstrip resonator circuit can be fabricated on a suitable ceramic substrate such as strontium titanate ($SrTiO_3$). A dc bias voltage is used to change the dielectric constant of the ceramic substrate to achieve desired resonant frequency change (Figure 5.3) and, consequently, the phase shift based on the change in EM-wave velocity as a function of permittivity. A phase shift greater than 15% is possible in suitable ferroelectric materials, but impedance matching presents a problem over a frequency bandwidth greater than 10% in a variable velocity transmission line. High-performance ceramic substrates with optimum responsiveness at higher cryogenic temperatures are most desirable for fabrication of these components.

5.3.2 Performance Capabilities and Limitations

The performance of an HHTSCF phase shifter is dependent on the ferroelectric properties of the substrate, DJI transmission line parameters, dc bias level, and

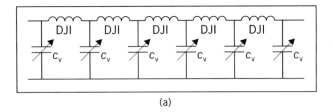

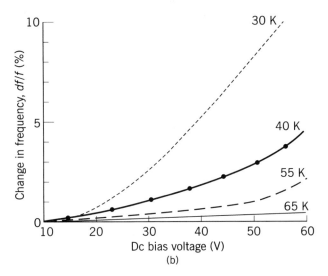

FIGURE 5.3 (a) Lumped element model of HTSC/ferroelectric phase shifter. (b) Frequency variation as a function of dc bias voltage and cryogenic temperature in a superconducting/ferroelectric phase shifter [2]. DJI = Distributed Josephson inductance; C_v = variable capacitance due to change in dielectric constant as a function of bias voltage v.

operating temperature. Maximum frequency shift and, thus, maximum phase shift are dependent on the optimum responsiveness of the ferroelectric ceramic substrate. Frequency shift or phase shift as a function of temperature and dc bias voltage is illustrated in Figure 5.3. It is evident that higher phase shift is possible at lower operating temperatures and higher bias voltage. Phase shift greater than 15% is possible with this device, which is more than adequate for most phased-array radars and communication systems.

5.4 PRECISION ANALOG SUPERCONDUCTING MICROWAVE (PASM) PHASE SHIFTER USING YBCO THIN FILMS

A PASM phased shifter is highly desirable where a precision control of beam steering and sidelobe level in electronically steerable phased-array antennas is required. In this phase shifter the magnetic field penetrates into the superconducting thin films, which changes the propagation constant of the superconducting microstrip transmission line. The amount of phase shift introduced depends on the characteristic impedance of the line, the length of the line, and the surface resistivity of the ground plane and microstrip transmission line [2].

5.4.1 Design Aspects and Performance Capabilities

The structural details of this phase shifter are shown in Figure 5.4. A 0.2-μm YBCO film with a critical temperature of 90 K is deposited on a cubic zirconia

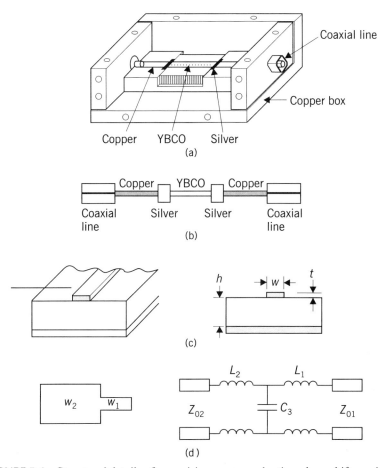

FIGURE 5.4 Structural details of a precision superconducting phase shifter using ferroelectric technology: (a) phase shifter assembly showing internal details; (b) interconnection of various phase-shifter elements; (c) microstrip transmission line details (w = width, t = thickness, h = substrate thickness); (d) equivalent electronic circuit representation.

substrate. A YBCO film as wide as the copper strip is soldered to the copper strip comprising silver film. The width of the silver film must be at least three times that of the copper strip width to ensure high mechanical integrity and reliability of the interconnection. The geometrical parameters must be optimized to obtain the best phase-shifter performance. This phase shifter offers precision phase control at microwave frequencies with low insertion loss, fine tuning, and fast tuning speed.

5.4.2 Mathematical Expressions to Compute Phase Shift

For a quasi-TEM transmission line with characteristic impedance Z_0 and propagation constant γ, the phase shift (ϕ), and the insertion loss (IL) can be expressed by the following equations:

$$IL = 10 \log S_{21}^2 \qquad (5.5)$$

$$S_{21} = [S_{21}]^2 e^{j\phi} \qquad (5.6)$$

where $S_{21} = 2Z_1/(AZ_1 + B + CZ_1^2 + DZ_1)$
 $[S_{21}]$ = absolute value of the parameter S_{21}
 Z_1 = load impedance
 $A = \cosh \gamma_1$
 $B = Z_0 \sinh \gamma_1$
 $C = (\sinh \gamma_1)/Z_0$
 $D = \cosh \gamma_1$
 $\gamma = \alpha + \beta$ = propagation constant, where α is the attenuation constant and β is the phase constant per unit length of the transmission line

5.5 HYBRID DESIGN OF A NONRECIPROCAL PHASE SHIFTER USING YIG-TOROID AND THE SUPERCONDUCTING TRANSMISSION LINE

5.5.1 Design Aspects and Performance Capabilities

Conventional phase shifters using microstrip technology have limited use at higher microwave frequencies because of high insertion loss due to high resistance of the normal metal conductor and poor unloaded Q of the transmission line. However, a phase shifter when fabricated from a YBCO superconducting microstrip transmission line on a lanthanum aluminate (LaAlO$_3$) coupled to a low-loss ferrite toroid offers a differential phase shifter greater than 700° with insertion loss not exceeding 0.7 dB [3]. This phase shifter offers superior performance because the magnetic flux required for the gyromagnetic interaction with the rf signal is contained within the closed magnetic path of the ferrite toroid. Specific details on the superconducting YBCO meanderline, YIG toroid, and magnetic biasing coil are shown in Figure 5.5, along with the coupling of the superconducting transmission line to the ferrite toroid. The YBCO meander-line circuit geometry provides the nonreciprocal phase shift parallel to the quarter-wavelength meander line at the center frequency. This provides circular polarization in the center of each meander section, which produces more phase shift for a given meander-line geometry. In a reciprocal phase shifter, the phase shift of the transmitted EM signal is independent of the direction of the magnetic polarization of the ferrite. In a nonreciprocal device, the transmission phase shift is dependent on the polarity of the magnetization of the ferrite, and, thus, the differential phase shift of this phase shifter is significantly improved.

The toroid establishes the collinearity of the magnetization with the meander-line segments for optimum performance. The differential phase shift can be changed by varying the dc current switching in the magnetic coil. A monolithic microwave structure can be used, in which the superconducting film is deposited directly on the ferrite. The superconducting YBCO film thickness must be compatible with London penetration depth, as shown in Figure 5.6. This planar ferrite substrate can provide the magnetic circuit, including the magnetic

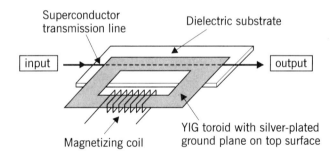

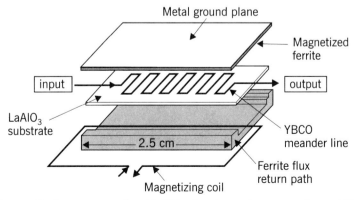

FIGURE 5.5 Design aspects of a 10-GHz, low-loss microwave phase shifter using superconducting thin film and ferrite [3].

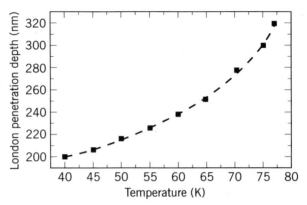

FIGURE 5.6 London penetration depth in a superconducting thin film (YBCO) as a function of cryogenic temperature (K).

path and the associated switching circuitry. A monolithic device design will simplify the packaging and will be most appopriate for high-T_c films when the technology for deposition of epitaxial superconducting films on ferromagnetic substrates is fully matured. A niobium circuit was deposited directly on a ferrite substrate and performance data were obtained at 4.2 K in 1994.

This phase shifter has an insertion loss less than 0.8 dB over a frequency range of 6–10 GHz, including the cable and coax-to-microstrip adaptor losses. Superconducting ohmic losses are expected to drop by a factor of 10 or more if the device operating temperature is reduced from 77 to 10 K. A peak differential phase shift greater than 700° has been demonstrated (Figure 5.7). For an optimized phase-shifter design configuration with ideal impedance matching of superconducting circuit and full utilization of saturation magnetization ($H_s = 4\pi M_s$), we may soon be able to achieve a figure-of-merit (FOM) value exceeding 5000°/dB.

5.5.2 Ferrite Requirements for a Superconducting, Hybrid Phase Shifter

The operating frequency for this type of phase shifter must be well above the resonance frequency of the ferrite material to achieve minimum insertion loss. The resonance frequency f_r is given by

$$f_r = g_m H_s \tag{5.8}$$

where g_m = gyromagnetic ratio (2.8 MHz/G, typical value)
 H_s = saturation magnetization of the ferrite
     = $4\pi M_s$

The value of 2.8 MHz/G is valid only at room temperature (300 K). For a ferrite material with H_s equal to 1220 G, the resonance frequency at room

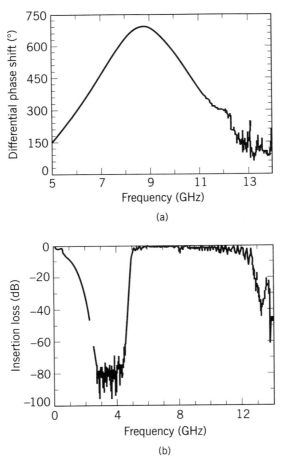

FIGURE 5.7 (a) Differential phase shift and (b) insertion loss of a superconducting phase shifter operating at 77K over 5–10 GHz [3].

temperature is about 3360 MHz. For a superconducting device, ferrite material with low Curie temperature must be selected. An industry survey of ferrite materials indicates that an aluminium-doped yttrium iron garnet (YIG) such as G-300 ferrite will be most suitable for superconducting ferrite devices, because its Curie temperature is less than 125°C. This is a temperature at which H_s approaches zero and the material becomes paramagnetic. The ferrite material G-300 has a saturation magnetization of 300 G at 300 K, which increases to 410 G at 77 K. In brief, a ferrite material with resonance frequency well below the operating frequency must be selected to avoid low-field insertion loss. A single-piece ferrite toroid must be used to avoid deterioration of the hysteresis loop squareness caused by the gap that can occur when two ferrite sections are joined to form a closed magnetic path. The operating frequency of this particular superconducting phase shifter can be easily extended to 30 GHz by selecting a ferrite with H_s greater than 5000 G and a Curie temperature less than 200 K. The benefits of a superconducting phase shifter would be even greater at MM-wave frequencies.

5.6 SUPERCONDUCTING MAGNETICALLY TUNABLE FILTERS USING FERRITES

In this section, magnetically tunable filters using superconducting technology are described, with particular emphasis on performance improvement. Magnetically tunable filters, including YIG filters and E-plane filters, are investigated for effective integration of superconducting technology. Both insertion loss and stop-band rejection will improve under cryogenic operating temperatures. At cryogenic temperatures, the saturation magnetization ($4\pi M_s$) values will increase dramatically and the ferrite and microwave structure losses can be virtually eliminated using high-performance superconductors. The maximum value of saturation magnetization for a MM-wave ferrite material is about 5000 G at 300 K, which increases rapidly to about 10,000 G at a cryogenic temperature of 4.2 K. Based on 5000 G value of a ferrite, the efficiency of the ferrite will begin to deteriorate beyond a 25-GHz frequency at room temperature because ferrite path lengths can no longer be shortened in proportion to operating wavelength by simply raising the value of $4\pi M_s$. This shortcoming can be overcome by operating the ferrite devices at cryogenic temperatures.

Superior filter performance at cryogenic temperature is possible with ferrites with narrow linewidths and low Curie temperatures. The nonlinear effects in ferrites under high rf power conditions will deteriorate the device performance because of internal temperature rise in the medium. These nonlinear effects can be eliminated under moderate input power levels at cryogenic operations. Preliminary performance calculations indicate that the combined microwave structure and ferrite loss can be reduced a factor of 3:1 and the reflection loss with optimum impedance matching by a factor of 2:1 at an operating temperature of 4.2 K.

5.6.1 Superconducting YIG-Tuned Bandpass Filter

This filter employs a well-polished, spherical YIG elements with narrow linewidth and low Curie temperature. The magnetic biasing field required to tune the filter over the band of interest is supplied by the tuning coil carrying dc current. The center frequency of the YIG filter varies with the biasing magnetic field intensity, which is a function of the dc drive current. The tuning speed is dependent on the time constant of the tuning coil. The tuning speed, insertion loss, and selectivity all improve with a decrease in operating temperature. The tuning speed improves because of the time constant improvement at low temperature. The insertion loss improves due to reduced ferrite losses. The conductor ohmic losses decrease due to substantial reduction in surface resistance. Selectivity is improved due to improvement in the quality factor of the microwave structure by several orders at cryogenic temperature. The projected performance improvement of a 2-stage YIG-tunable filter at cryogenic temperatures is summarized in Table 5.2.

The values in Table 5.2 are projected values for insertion loss and stop-band rejection or isolation for a 2-stage YIG-ferrite tunable filter as a function of frequency and operating temperature. One can expect considerable improvement in these parameters at 4.2 K. The ferrite used was not an optimum material with low Curie temperature. An aluminium-doped YIG ferrite such as G-300 with a Curie temperature of 120°C will yield much better performance.

5.6.2 Magnetically Tunable E-Plane Filter Using Superconducting Technology

The power-handling capability of a YIG-tuned filter is limited to a maximum of +20 dBm due to the limiting problem produced by the nonlinear characteristic of the material at higher power levels. Structural details of an E-plane filter are shown in Figure 5.8. An E-plane magnetically tunable filter (MTF) offers insertion loss as low as 1.5 dB and cw power-handling capability as high as 2 W at room temperature in the Ku-band region. The room temperature (300 K) insertion loss can be further reduced to 0.8 dB by loading the resonator on one side only. The external magnetic field is provided by a single wire coil around the ferrite YIG disks carrying the dc biasing current. The waveguide resonators are filled with lateral ferrite thin slabs of low-loss ferrite material with low Curie temperature. True MMIC compatibility is not possible with this filter configuration, but its potential application to high-power phased-array radar systems cannot be ruled out.

Preliminary calculations reveal that a 16-GHz, E-plane filter will have an insertion loss of about 3.5 dB at 300 K, 2.1 dB at 77 K, and 0.8 db at 4.2 K. Furthermore, the stop-band rejection of 45 dB will improve by a minimum of 8 dB and the cw power handling capability of 2 W will be increased to 4 W at 4.2 K operation. Integrated circuit technology, if used in the fabrication of this

TABLE 5.2 Performance Projection of a Superconducting YIG Filter

Frequency (GHz)	300 K		200 K		77 K	
	IL (dB)	Iso (db)	IL (dB)	Iso (db)	IL (dB)	Iso (db)
8	3.5	48	2.5	51	1.2	53
12	4.0	47	2.7	49	1.3	51
16	4.3	46	2.8	48	1.6	49
18	4.5	45	2.9	47	1.7	47

Note. IL, insertion loss; ISO, isolation.

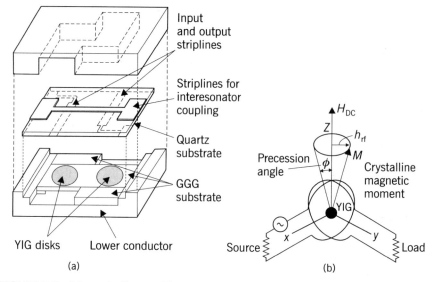

FIGURE 5.8 Magnetically tunable E-plane microwave bandpass filter using YIG and superconductor technologies: (a) internal details of the filter; (b) interaction between the applied magnetic field and rf signal. Rf tuning range = 14.1–15.7 GHz, passband insertion loss = 3.5 dB at 300 K (max.) and 2.1 dB at 77 K (est.), and stopband rejection = 45 dB (min.). The tuning coil around the ferrite YIG disks carrying bias current is not shown.

filter, can offer a low-cost photolithographic fabrication technique that is ideal for MM-wave superconducting ferrite tunable filters. Note TransTech ferrite materials TTI-2800 and TTVG-1200 are best suited for MM-wave magnetically tunable filters.

5.6.3 Superconducting Ferrite Circulators and Isolators

Even the passive ferrite devices such as circulators and isolators will benefit in terms of insertion loss, isolation, and voltage standing wave ratio (VSWR) if the operating temperature is maintained around 77 K. Substantial reduction in insertion loss, particularly at MM-wave frequencies, can be realized by selecting the optimum ferrite material and cryogenic temperature. High-power switching circulators, which are considered active components, will benefit the most from superconducting technology. Performance improvement in an X-band circulator operating over 8150–8750 MHz as a function of cryogenic temperatures can be seen from the data summarized in Table 5.3. Both the power-handling capability and the limiting threshold level of a device improve at cryogenic temperature, in addition to improvements in insertion loss and isolation.

5.6 SUPERCONDUCTING MAGNETICALLY TUNABLE FILTERS USING FERRITES

TABLE 5.3 Projected Performance Data on an X-Band Circulator at Cryogenic Temperatures

Temperature (K)	Insertion Loss (dB)	Isolation (dB)	VSWR
300	0.32	19	1.30
100	0.22	21	1.20
77	0.17	26	1.15
4.2	0.10	32	1.10

5.6.4 Ferrite Material Requirements for Superconducting Circulators and Isolators

Ferrites are nonreciprocal materials characterized by a Hermitian permeability tensor. When the static field is nonuniform, the local coordinate system rotates into the alignment with the static magnetization. The transverse magnetic field components, which are coupled by the off-diagonal tensor element, vary throughout the ferrite medium. The tensor elements are given by

$$\mu_t = \frac{1 + \omega_0 \omega_m}{\omega_0^2 - \omega^2} = [1 + k] \quad (5.10)$$

$$k = \frac{\omega \omega_m}{\omega_0^2 - \omega^2} \quad (5.11)$$

$$\omega_0 = -g_m \mu_0 H_0 \quad (5.12)$$

$$\omega = -g_m \mu_0 M_s \quad (5.13)$$

where g_m = gyromagnetic ratio of the ferrite
H_0 = internal static bias magnetic field in the ferrite
M_s = saturation magnetization of the ferrite material
μ_t = tangential component of the complex permeability
$\omega_0, \omega_m, \omega$ = appropriate angular frequencies

The ferrites are dispersive, with the natural procession of the magnetic moments around the magnetic bias field. The frequency of this procession (Figure 5.8) is determined by the magnetic bias field intensity. The components of the complex permeability tensor approach infinity when this procession undergoes resonance, which is known as ferromagnetic resonance. The permeability tensor is strictly dependent on the permeability, saturation magnetization, and gyromagnetic ratio of the ferrite material.

The magnetic losses are included in the permeability tensor and are a function of linewidth and permeability. The rf or electrical losses are included in the complex permittivity of the ferrite and are strictly dependent on the dielectric constant and loss tangent of the material used. The microwave structural losses are dependent on the surface resistance of the circuit. Both the ferrite losses and microwave circuit losses are functions of operating temperature and their magnitudes will decrease under cryogenic temperatures, particularly, in ferrites with low Curie temperatures.

The gyromagnetic ratio in most undoped crystalline ferrites remains unchanged at cryogenic temperatures, except in some specific doped ferrite materials. The saturation magnetization increases with the decrease in operating temperature. At cryogenic temperatures, change occurs in the ferrite crystal structure, resulting in the conversion of divalent iron ions (Fe^{2+}) into trivalent iron ions (Fe^{3+}) within the octahedral position which causes constant change in saturation magnetization.

Variation in linewidth as a function of cryogenic temperature is not yet fully known. However, it is evident from Figure 5.9 that saturation magnetization for most ferrites increases as the operating temperature is reduced. This will push the operating frequency well above the ferrite resonance frequency, where the insertion loss is minimum in the ferrite medium. The resonance frequency, which is proportional to saturation magnetization, will go up. This will require an operating frequency greater than that defined by Kittel equation. The

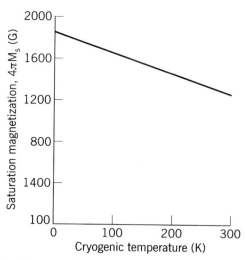

FIGURE 5.9 Variation in saturation magnetization in aluminum-doped yttrium iron garnet (YIG) as a function of cryogenic temperature. Room temperature properties of TRANS-TECH ferrite G-1200: Saturation magnetization = 1200 G, 3-dB linewidth at 10 GHz = 40 Oe, dielectric constant at 10 GHz = 14.8, loss tangent at 10 GHz = < 0.0002, Curie temperature = 220°C.

applied magnetic field (H_a) based on Kittel theory for achieving resonance increases as the operating temperature is reduced, which is evident from Equation (5.13). The loss in ferrite are small at internal static field above saturation and about $5\Delta H$ below gyromagnetic resonance. Note that ΔH is the width of peak magnetic field at half-amplitude at ferromagnetic resonance. The applied magnetic field H_a is defined as

$$H_a = \frac{f_0}{g_m} + 4\pi M_s \tag{5.14}$$

where f_0 is the operating frequency in megahertz, and the constant g_m is known as the gyromagnetic ratio whose room temperature value is 2.8 MHz/Oe.

The permeability of the ferrite decreases with the decrease in operating temperature. The losses in the ferrite are directly proportional to dielectric constant (e_r) and the loss tangent (tan δ) of the material. The dielectric constant of ferrites remain unchanged up to its Curie temperature; the material becomes paramagnetic as the operating temperature approaches the Curie temperature.

5.6.5 Impact of Cryogenic Operation on Ferrite Losses

Ferrite losses are dependent on the dielectric constant and loss tangent. The dielectric constant remains unchanged even under cryogenic operations in most ferrites. But the loss tangent is proportional to the permeability of the materials, which undergoes considerable reduction at cryogenic temperatures, as shown in Table 5.4. This means that the permeability and, thus, the insertion loss in the ferrite decrease with operating temperature. The data shown in Table 5.4 indicate that the initial room temperature (300 K) permeability (μ_0) is reduced by a factor of 5:1 at 77 K in both MnZn and ZnNi ferrite materials. One can expect further reduction in permeability and loss tangent and, thus, in ferrite losses if the operating temperature is reduced to 4.2 K. It is clear from the data in Table 5.4 that the losses in the ZnNi ferrite will be minimum at 77 K compared to those in MnZn ferrite.

At cryogenic temperatures, the anisotropy constant K_1/M and K_2/M undergo drastic changes in both sign and magnitude. In the case of $Mn_{0.96}Co_{0.04}Fe_2O_4$ ferrite, both the anisotropy constants undergo drastic change at cryogenic temperatures. The ratio K_1/K_2 varies between 2 and 10

TABLE 5.4 Permeability as a Function of Cryogenic Temperature for Various Ferrites

Temperature (K)	MnZn Ferrite	ZnNi Ferrite
300	1350	265
175	1025	135
77	255	53

for this particular material over the temperature range of 300 to 4.2 K. Variations in the anisotropy constants for these two ferrites as a function of operating temperature are shown in Table 5.5. The effects of variation in anisotropy constants will impact the performance of phase shifters, isolators, and gyrotrons operating at cryogenic temperatures. How the variations in these constants will effect a specific device performance parameter is not fully known. Studies performed on microwave ferrites in the past reveal that zero anisotropy in a ferromagnetic material is highly desirable for satisfactory operation over wide frequency bandwidth.

5.6.6 Impact of Cryogenic Temperature on Anisotropic Field and Linewidth of YIG Materials

The magnetic loss in a ferrite device is directly proportional to the linewidth of the material. The linewidths of doped YIGs at 24 GHz and undoped YIGs at 10 GHz as a function of cryogenic temperature are shown in Figure 5.10. Large line-broadening effects due to impurities are evident. Narrow linewidth means that low magnetic loss is possible only at lower operating temperatures, regardless of whether the YIG ferrite is doped or undoped. The anisotropic field increases as the operating temperature is reduced, as shown in Figure 5.10b.

Recent advances in ferrite material technology indicate that spinel ferrites are the most desirable for MM-wave ferrite devices. Isolation and insertion loss parameters for various MM-wave isolators using spinel ferrites as a function of operating temperature are shown in Table 5.6. Considerable improvement in device performance parameters can be seen at cryogenic temperatures. The

TABLE 5.5 Impact of Cryogenic Operation on Anisotropy Constants

Temperature (K)	K_1/M	K_2/M	K_2/K_1
300	−20	−50	2.5
260	+21	−105	5.0
240	+58	−150	2.7
77	+2330	−7100	3.0
4.2	> 0	< 0	—

TABLE 5.6 Performance Improvement in Superconducting MM-Wave Isolators Using Spinel Ferrites

Frequency (GHz)	Insertion Loss (dB)		Isolation (dB)	
	300 K	77 K	300 K	77 K
33–50	0.50	0.25	25	29
40–60	0.52	0.30	25	28

5.6 SUPERCONDUCTING MAGNETICALLY TUNABLE FILTERS USING FERRITES

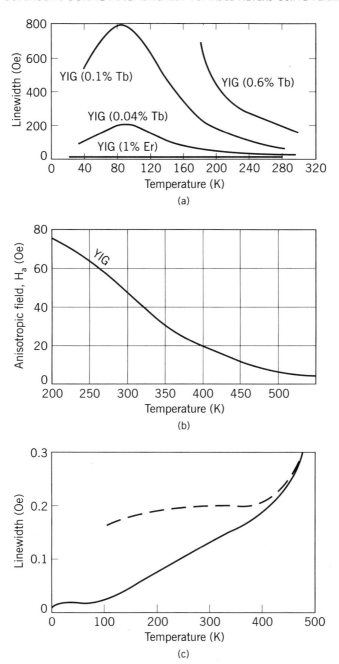

FIGURE 5.10 Linewidth and anisotropic field for various ferrites as a function of cryogenic temperature. (a) Linewidth of various doped YIGs at 24 GHz (Tb, terbium; Er, erbium). (b) Anisotropic field, H_a. (c) Linewidth for undoped YIG ferrites at 10 GHz (– – –, best commercial YIG; —, standard YIG).

insertion loss in an isolator operating over 18 to 23 GHz varies from 1.20 dB at 300 K to 0.20 dB at 77 K to 0.06 dB (projected) at 4.2 K. The improvement in isolation varies from 20 to 26 dB over the same temperature and frequency range. The values of isolation and insertion loss are limited to an rf bandwidth of 0.5 GHz in the frequency ranges specified. The data clearly indicates a performance improvement of the spinel ferrite devices at cryogenic temperatures. Further improvement in both insertion loss and isolation can be expected at 4.2-K operation.

5.6.7 Impact of Cryogenic Temperature on Resonance Frequency

The precessional frequency or the resonance frequency f_r of a ferrite device is proportional to the bias magnetic field H_0 and the anisotropic field H_a. However, the anisotropic field varies as function of temperature, as shown in Figure 5.10. This means that a slight change in the resonance frequency of a ferrite device can be expected as a function of operating temperature. The resonance frequency of a ferrite device is given as

$$f_r = g_m(H_0 + H_a) \qquad (5.15)$$

where g_m is the gyromagnetic ratio of the electron (2.8 MHz/Oe).

5.6.8 Effective Gyromagnetic Ratio (g_{eff}) at Cryogenic Operations

The gyromagnetic ratio is strictly dependent on the dopant fractional amount and operating temperature. For example, in a nickel acuminate ferrite (NiFe$_x$Al$_y$O$_4$ where $x = 1 - y$), the effective gyromagnetic ratio is 6.6 when y is equal to 0.6 and is reduced to 2.2 when y is equal to zero at 77-K operating temperature. This means that the bias magnetic field requirement is reduced by a factor of 3:1 at 77-K operation, which can reduce the weight and size of the device magnet by a factor of 9:1, because the weight of the magnet is proportional to the square of the bias field. Further reduction in the bias magnetic field and consequently in device magnet weight and size is expected at 4.2-K operation. No significant change in anisotropic field is anticipated at this cryogenic temperature.

5.7 CRYOGENIC WINDOWS FOR HIGH-POWER MM-WAVE SOURCES

High-power MM-wave sources, namely, gyrotrons, gyro-TWTAs, and klystrons, employ cryogenic windows to achieve high peak power levels at MM-wave frequencies with high conversion efficiencies. Gyrotrons are widely used for plasma heating. Plasma heating requires high peak power with high duty cycles to achieve quasi-cw operation. Extremely long pulses are used in high-power gyrotrons and all the elements are designed for cw operation.

5.7 CRYOGENIC WINDOWS FOR HIGH-POWER MM-WAVE SOURCES

Cryogenically cooled windows and matching optics are used in high-power gyrotrons. A 118-GHz, 500-kW, gyrotron with a liquid nitrogen (77-K) edge-cooled window was designed and developed by French and Swiss scientists. The sapphire disk window has a diameter of 12 cm and is 0.625 cm thick. The computed runway power is 650 kW, which provided enough of a safety margin for operation at 500 kW to be delivered into a high-power load. The electronic efficiency was 33% and the overall efficiency was 28% with a cryogenically cooled sapphire window.

Electron cyclotron wave (ECW) systems are highly desirable for plasma heating and nuclear fusion devices. They require gyrotrons with cw power capability exceeding 1 MW in the frequency range of 140–170 GHz. A safe ECW window is one of the most challenging ones. The rf window can be made from sapphire or high-grade diamond or high-quality silicon with high resistivity. A cryogenically cooled window not only provides high cw power levels with improved efficiency, but also offers some protection from radiation damage from X-rays. Preliminary calculations on silicon-based windows indicate that these windows are capable of transmitting 1 MW of power over 140–200 GHz frequency by edge cooling to maintain a window temperature of 260 K. This is the best performance of a cryogenically cooled silicon window reported to date, because the dielectric losses at 145 GHz are the lowest at 260-K window temperature. The loss tangent of this silicon window is less than 0.000015 at 260 K and at 145-GHz frequency.

The absorption coefficient and loss tangent of a cryogenically cooled window are the most critical performance parameters, because they have substantial impact on the conversion efficiency, reliability, and longevity of the rf window as well as the MM-wave source. Cost-effective studies performed in the past indicate that sapphire material for the high-power rf window is most suitable. The sapphire windows are widely used in high-power MM-wave sources such as gyrotron, klystron, gyroklystron, gyro-traveling-wave amplifier (gyro-TWA), and gyro-backward wave oscillator (gyro-BWO). The absorption coefficient and loss tangent for a sapphire rf window as a function of cryogenic temperature and operating frequency are shown in Figure 5.11.

Mechanical strength and thermal consideration were not adequately addressed when the sapphire windows were initially employed in high-power MM-wave sources operating beyond 110 GHz. Published data [4] on window materials indicate that the increase in absorption coefficient with increasing MM-wave frequencies poses a serious problem for the high-power rf windows. A crystalline sapphire material offers high mechanical strength and reasonably high thermal conductivity, but suffers from a high absorption coefficient, particularly at frequencies above 110 GHz, even at 4.2 K operation. Several studies on high-power rf window materials recommend diamond-like materials such as single-crystal silicon with high resistivity and natural diamond. When a wafer of silicon is coated on both sides of a synthetic diamond film, considerable improvement is possible in both the mechanical strength and thermal conductivity with improved absorption coefficient at cryogenic operations.

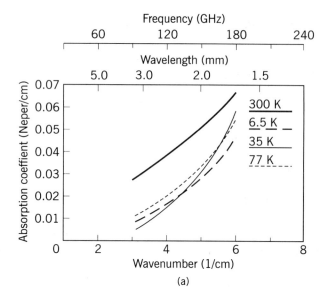

FIGURE 5.11 (a) Absorption coefficient and (b) loss tangent for a high-power sapphire window at cryogenic temperatures [4].

These cryogenically cooled windows are best suited for high-power gyrotrons operating in the 100- to 280-GHz range and beyond. The heat generated in the rf window depends on the dielectric loss, which is a function of permittivity of the material and the operating temperature. Permittivity of diamond and sapphire windows as a function of cryogenic temperature at 145 GHz is depicted in Figure 5.12. The loss tangent for various cryogenically cooled windows is shown in Figure 5.13. It is evident from Figure 5.13 that significant reduction in the loss tangent for silicon and diamond windows occurs around 265 K.

5.7 CRYOGENIC WINDOWS FOR HIGH-POWER MM-WAVE SOURCES

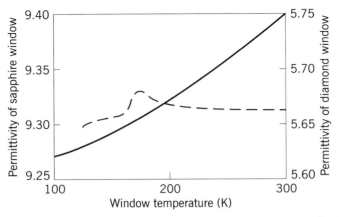

FIGURE 5.12 Permittivity of diamond (– – –) and sapphire (—) rf windows at 145 GHz as a function of cryogenic temperature [4].

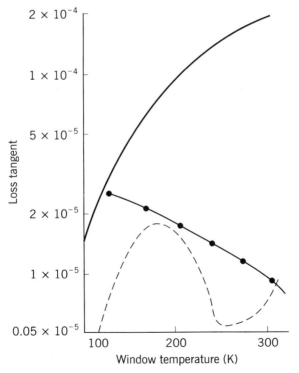

FIGURE 5.13 Loss tangent for sapphire (—), diamond (–•–), and silicon (– – –) windows as a function of cryogenic temperature [4].

5.8 SUPERCONDUCTING MICROWAVE SWITCH FOR CONTROL APPLICATIONS

Low-loss microwave switches are highly desirable for high-performance control applications. Conventional rf switches using MESFET or PIN diode solid-state devices suffer from high insertion losses. A microwave superconducting switch using an HTSC material such as TlCaBaCuO embedded in a microstrip transmission line can offer insertion loss of less than 1 dB over the 17:1 bandwidth ratio. This switch development work was supported by the U.S. Department of Energy in 1990. The superconducting switch demonstrated an insertion loss of less than 1 dB, isolation better than over 30 dB, and switching speed less than 1 µs over 0.5–8.5 GHz. A switching speed of less than 1 µs is more than adequate for many microwave and millimeter-wave (MM-wave) applications, such as switched-line phase shifters, variable electronic attenuators, and leveling circuits in sweep generators.

5.8.1 Design Aspects and Structural Details

The design of the thermal HTSC switch is based on driving a small thin superconducting bridge into the normal state. The bridge can be embedded in a 50-Ω coplanar waveguide. The TlCaBaCuO base film is patterned with standard optical lithography on a lanthanum aluminate ($LaAlO_3$) substrate. The bridge region dimensions can be 200 by 10 µm. The bridge thickness is about 50 nm and the transmission line thickness is 300 nm. This reduces the power requirements for driving the bridge normal by reducing the total heat load in the active area. A thermal insulator is placed over the control line in the switch area to further reduce power requirements.

5.8.2 Performance Capabilities

The control current can be selected for quick turn-off and turn-on times. The control current produces just enough heat to drive the bridge normal but does not raise the local temperature exceedingly high. The power dissipation in the control line will be less than 5 mw at 77 K operation, which can be further reduced at 4.2 K. No performance variation or degradation is anticipated over input power ranging from 0 to −50 dBm.

This type of switch can be easily integrated with printed circuit transmission lines, microstrip transmission lines, coplanar waveguides, and a variety of other transmission systems with minimum cost and complexity. The design offers MMIC compatibility for many microwave devices. Potential applications include high-performance signal control circuits, switched delay line phase shifters for phased-array radar systems and signal distribution circuits.

The switching time of 1 µs is limited by the thermal time constant. Many control applications do not require a switching speed faster than 1 µs. Faster switches will require the use of Josephson junctions or flux flow devices. Both switch types suffer from poor isolation and complicated circuitry. Furthermore, semiconductor-based rf switches are more difficult to integrate with HTSC materials and have insertion losses typically on the order of 2 dB at X-band frequencies. As stated earlier, this particular superconducting switch demonstrated an insertion loss of less than 1 dB and isolation better than 30 dB at 77-K operating temperature over a frequency spectrum of 0.5–8.5 GHz. Improved impedance matching, optimum microwave structure design, and lower cryogenic temperature can reduce the insertion loss to less than 0.5 dB and increase the isolation to greater than 40 dB at X-band frequencies. The performance of this switch is much better than that of conventional rf switches using GaAs solid-state devices. This device is a good choice for large phased-array radar systems, where cost is of prime consideration.

SUMMARY

Integration of superconducting technology in various microwave and MM-wave components has significantly improved their performance. In the case of superconducting phase shifters, remarkable improvement in insertion loss and switching speed has been reported. Conventional bandpass filters and YIG-based tunable filters have demonstrated significant reduction in insertion loss at cryogenic temperatures. For ferrite circulators and isolators, cryogenic operation demonstrated significant improvement in terms of insertion loss and isolation. Implementation of HTSC technology in high-power microwave windows is necessary for reliable operation of electron cyclotron systems, high-power klystrons transmitters, plasma heating equipment, and nuclear fusion systems. Cryogenic operation of high-power microwave windows offers optimum safety, reliability, longevity, and uninterrupted operation over extended periods.

REFERENCES

1. G. M. Jackson et al. A high-temperature superconductor phase shifter. *Microwave J.*, 72–78 (December 1992).
2. S. Ca and D. Chef. High-T_c superconducting microwave phase shifter. *J. I MM-Wave* 15:439–449 (1994).
3. G. Dionne et al. YBCO/ferrite low-loss, microwave phase shifter. *IEEE Trans. Appl. Superconductivity* 5:2083–2087 (1995).
4. M. N. Afsar and K. J. Button. MM-wave dielectric properties of materials. *IEEE Proc.* 73:131–151 (1985).

CHAPTER SIX

Performance Improvement of Solid-State Devices at Cryogenic Temperatures

This chapter focuses on the performance improvements of microwave and MM-wave solid-state devices at cryogenic temperatures. The solid-state devices include PIN diodes, silicon and GaAs varactor diodes, Schottky barrier mixer diodes, GUNN diodes, metal-semiconductor field effect transistors (MESFETs), modulation-doped FETs (MODFETs), which are also known as two-dimensional electron gas FETs (2-DEGFETs), high electron mobility transistors (HEMTs), pseudomorphic-HEMTs (p-HEMTs), doped-channel heterojunction FETs (DC-HFETs), inverted-HEMTs (i-HEMTs), superconductor-base transistors (SBTs), which are also called flux flow transistors, and heterojunction bipolar transistors (HBTs). The impact of cryogenic temperatures on the characteristics of various two-terminal and three-terminal solid-state devices is discussed in detail. Microwave and MM-wave solid-state components and circuits using these devices are identified wherever possible, but these components and circuits are discussed in detail in Chapter 7. The discussion in this chapter is limited to those microwave and MM-wave components and circuits that benefit most from the integration of superconducting technology. Cryogenically cooled components such as GaAs power amplifiers, low-noise amplifiers (LNAs), solid-state oscillators, parametric amplifiers, frequency multipliers, mixers, and filters are considered for possible application to radar systems, satellite communication systems, electronic warfare (EW) equipment, and space sensors.

6.1 MICROWAVE DIODES

Performance capabilities are described of some specific microwave diodes—PIN diodes, varactor diodes, GUNN diodes, and Schottky barrier diodes—

operated at superconducting temperatures. Quantum-well (QW) diodes are discussed along with their potential applications to electro-optic components and systems. Figure of merits of cryogenic-cooled varactor diodes are specified wherever possible. Improvement in quality factors of various diodes at cryogenic operations is identified.

6.2 VARACTOR DIODES

Varactor diodes are widely used in voltage-tunable oscillators, frequency multipliers, up and down converters, and parametric amplifiers. The varactor is a solid-state device whose junction capacitance is a nonlinear function of reverse bias voltage across the junction (Figure 6.1). The parameters are defined as follows:

L_p = varactor package capacitance
C_j = nonlinear junction capacitance as a function of bias voltage
C_p = package capacitance
R_s = overall series resistance of the device, which is a sum of substrate resistance (R_{sub}), contact resistance (R_{con}), bond wire resistance (R_{wire}), and undetected epitaxial layer resistance (R_{epi}).

For a perfect varactor diode with single junction, R_{epi} is the only contributing element to the series resistance R_s. The contact resistance has two components: top and bottom contact resistance, both of which decrease with the decrease in operating temperature. A typical value of contact resistance is about 0.1 Ω at 300 K, which reduces to about 0.01 Ω at 77 K.

The bonding wire resistor is made from high electrical conductivity material. Its room temperature (300 K) value is reduced to less than 10% at 77 K. The substrate acts like a capacitor in which the copper loss is theoretically is zero. Furthermore, the resistive losses in package capacitance and inductance are essentially zero. This leaves only the epitaxial layer resistance, which is the major contributor of the R_s parameter. Its value varies from 0.5 to 1.0 Ω at 300 K for a perfect varactor.

The quality factor is proportional to cutoff frequency (f_c) of the diode and is defined as follows:

$$Q = \frac{F_c}{F_o} \quad (6.1)$$

where $F_c = \dfrac{1}{2\pi R_s C_{jmin}}$ (6.2)

F_o = operating frequency
C_{jmin} = minimum junction capacitance.

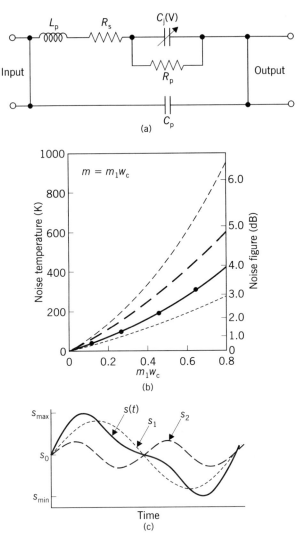

FIGURE 6.1 (a) Equivalent circuit, (b) performance parameters (– – –, 300 K; — — —, 200 K; –●–, 100 K; - - -, 77 K, and (c) elastance curves for a varactor noise figure of a parametric amplifier.

The junction capacitance is usually less than 0.2 pF and remains unaffected by the operating temperature. The series resistance is the only parameter that varies as a function of operating temperature.

6.2.1 Improvement in Diode Insertion Loss

The insertion loss in a varactor is strictly dependent on its series resistance. For a high-performance GaAs hyperabrupt junction varactor diode the series resistance varies from roughly 1 Ω at 300 K to 0.14 Ω at 77 K. Thus, the insertion loss or the copper loss is about seven times more at 300 K than at 77 K operating temperature.

6.2.2 Varactor Leakage

Hyperabrupt varactor junction (HAJ) diodes are best suited for voltage-controlled oscillators (VCOs), which are widely used in missile seekers, EW receivers, and telecommunication systems. Linear tuning, leakage current, and post-turning drift are the most important parameters of a VCO, but leakage current is the most critical. Preliminary calculations indicate that silicon HAJ tuning varactors have leakage currents of about 20 pA at 300 K, 1 pA at 200 K, 0.03 pA at 100 K, and 0.002 pA at 77 K. Uniform linear capacitance–voltage (C-V) characteristic over temperature and tuning period is absolutely necessary. Furthermore, the package inductance, tuning efficiency, linearity, phase noise, and leakage current decrease drastically as the tuning diode temperature is reduced well below 77 K.

6.2.3 Varactor Noise Figure

Varactors are best suited for cryogenically cooled, microwave parametric amplifiers, because they offer minimum noise figure under cryogenic operating temperatures. The noise figure of a parametric amplifier is dependent on the elastance of the varactor, the cutoff frequency of the diode, and the modulation ratio. The cutoff frequency is the most dominant parameter, and is inversely proportional to the diode series resistance. Series resistance is reduced by a factor of 7:1 when the operating temperature is dropped from 300 K to 77 K. As a result of improvement in the cutoff frequency of the varactor at cryogenic temperatures, the noise figure of the varactor is significantly reduced.

6.2.4 Thermal Performance and Power Dissipation

GaAs varactors are widely used in voltage-controlled oscillators, frequency multipliers, parametric amplifiers, and up/down converters. Thermal performance of these components in high-power systems is strictly dependent on the thermal conductivity of the diode junction material. Temperature-dependent thermal conductivity $K(T)$ plays a key role in heat-transfer efficiency. The thermal conductivity parameter for a GaAs varactor diode is expressed as follows:

$$K(T) = \frac{0.44}{(T/300)^{1.25}} \text{ W/cm}\,^\circ\text{C}^{-1} \tag{6.3}$$

where T is the diode junction operating temperature in kelvins. The calculated value of this parameter varies from 0.44 at 300 K to 0.73 at 200 K to 1.73 at

100 K to 2.41 at 77 K. Both heat-transfer efficiency and power dissipation improve at cryogenic operations. When the GaAs varactor diode is mounted on a copper heat sink, its thermal performance is further improved. Thermal conductivity of a GaAs varactor is calculated to be roughly 3.98 W/cm K^{-1} at 300 K, 4.13 at 200 K, 4.83 at 100 K, 6.22 at 77 K, 105 at 20 K, and 196 at 10 K operating temperature.

6.2.5 Improvement in Diode Cutoff Frequency and Quality Factor

Equation 6.1 shows that the cutoff frequency of a varactor is inversely proportional to its series resistance. Since the series resistance (R_s) decreases rapidly at cryogenic temperatures, the cutoff frequency increases rapidly as the operating temperature is reduced. The quality factor of the varactor is defined as the ratio of cutoff frequency to operating frequency ($Q = F_c/F_o$). Calculated values of cutoff frequency and quality factor at 10 GHz as a function of temperature are summarized in Table 6.1

We see that the figure of merit (FOM), which is the performance indicator of a varactor, improves at cryogenic temperatures. Improvement in quality factor will increase the cutoff frequency of the diode. Furthermore, improvement in the varactor Q will improve the frequency stability in a varactor-tuned VCO, the conversion efficiency in a frequency multiplier, and the gain and noise figure in a parametric amplifier. Performance improve in these parameters at cryogenic temperatures has been observed, regardless of whether a varactor is a mesa junction device, an abrupt junction device, or a hyperabrupt device. Significant improvement in cutoff frequency is observed in GaAs varactors compared to silicon varactors. The varactor FOM is also dependent on the electron density of the material, which is a function of temperature. In a Si varactor, its value of 9×10^{14} remains constant over the 500 to 125-K range which is reduced to 6×10^{11} at 20 K.

TABLE 6.1 Calculated Values of Cutoff Frequency and Series Resistance as a Function of Operating Temperature

Temperature (K)	$T^{1.5}$	R_s (Ω)	F_c (GHz)	Q at 10 GHz
300	5196	1.00	800	80
200	2828	0.56	1430	143
100	1000	0.21	3080	308
77	675	0.14	5174	517

6.3 PIN DIODES

PIN diodes are widely employed in microwave switches, attenuators, phase shifters, limiters, and other rf components and circuits. Series resistance and power dissipation are significantly reduced when the PIN diode junction temperature is maintained at 77 K or less. When the power dissipation in a diode is reduced, the power output or the power-handling capability increases, which is evident from the following equations.

6.3.1 Power Dissipation

$$P_{out} = P_{in} - P_{diss} \tag{6.4}$$

where P_{out} = output power
P_{in} = input power
P_{diss} = power dissipated in junction, which is defined as

$$P_{diss} = 150 - \frac{T_c}{R_{th}} \text{ for GaAs PIN diode} \tag{6.5}$$

$$= 175 - \frac{T_c}{R_{th}} \text{ for Si PIN diode} \tag{6.6}$$

T_c = CHIP temperature (°C)
R_{th} = thermal resistance (°C/W)

Thermal resistance increases with the decrease in the operating temperature, which will reduce the power dissipation at cryogenic temperatures. Thus, higher output power levels can be achieved by keeping the PIN-diode operating temperature at 77 K or lower. Power dissipation is essentially is defined as $I_f^2 R_s$ loss in the series resistance R_s, where I_f is the forward bias current. The series resistance of a PIN diode is defined as

$$\left[\frac{W_{int}}{\mu_e + \mu_h}\right] \cdot \left[\frac{1}{Q}\right] \tag{6.7}$$

where W_{int} = intrinsic region width (cm)
μ_e = electron mobility (6800 cm^2/V · s at 300 K)
μ_h = hole mobility (680 cm^2/V · s at 300 K)
$Q = I_f/\tau_c$ = forward current to carrier lifetime ratio

Both mobilities increase with a decrease in operating junction temperature, but the electron mobility increases faster than the hole mobility. This means that both the series resistance and the associated insertion loss in a PIN diode decrease as the PIN-diode junction temperature is reduced. Calculated values of series resistance and the associated loss at cryogenic temperatures are shown in Table 6.2. The assumed values of series resistance take into account the

TABLE 6.2 Impact of Cryogenic Operations on PIN Diode Performance

Temperature (K)	R_s (Ω)	Insertion Loss (dB)
300	1.00	0.172
200	0.56	0.097
100	0.21	0.035
77	0.14	0.024

magnitudes of the mobilities shown in Equation (6.7) as a function of temperature, which involve a transmission line with characteristic impedance of 50 Ω.

6.4 GUNN DIODES

Both the GaAs and silicon GUNN diodes are widely used in high-power MM-wave solid-state sources. Significant improvement in frequency stability, phase noise, conversion efficiency, and output power is possible at cryogenic temperatures. This improvement is due to reduction in series resistance, improvement in thermal performance, and enhancement in the FOM of the GUNN diode at cryogenic operations. The thermal conductivity will increase as the diode temperature is reduced below 77 K, which will reduce the power dissipation in the diode junction. Preliminary calculations indicate that output power of GUNN diodes at MM-wave frequencies increases rapidly as the operating temperature is reduced well below 77 K. Calculated values of power output from GaAs GUNN diodes as a function of frequency and operating temperatures are shown in Table 6.3.

6.5 IMPROVED RELIABILITY OF DIODES AT CRYOGENIC TEMPERATURES

The reliability of semiconductor microwave diodes such as GaAs Schottky barrier diodes, Si single-drift-region (SDR) diodes, IMPATT diodes, GUNN diodes, and PN-junction varactor diodes improves as the operating tempera-

TABLE 6.3 Power Output of a GaAs GUNN Diode as a Function of Frequency and Operating Temperature (mw)

Frequency (GHz)	300 K	77 K	4.2 K
35	175	225	284
77	65	89	145
94	56	78	95

ture is reduced to 200 K or below. The following expressions can be used to compte the mean time between failure (MTBF) for various microwave and MM-wave diodes at cryogenic operations:

$$\text{MTBF} = \frac{4.96}{10^{10}} \exp\left[(1.342)\frac{10^4}{T_j}\right] \quad \text{for GaAs Schottky diode} \quad (6.8)$$

$$= \frac{3.09}{10^{10}} \exp\left[(1.67)\frac{10^4}{T_j}\right] \quad \text{for Si-SDR IMPATT diode} \quad (6.9)$$

$$= \frac{7.72}{10^7} \exp\left[(1.18)\frac{10^4}{T_j}\right] \quad \text{for GaAs varactor diode} \quad (6.10)$$

Calculated values of MTBF for various diodes as a function of operating temperatures are shown in Table 6.4. The MTBF data indicate an improvement of several orders in the reliability of semiconductor diodes at 200 K. Further improvement in reliability is possible at lower cryogenic temperatures. This improvement in the reliability is strictly due to higher values of thermal conductivities of the electronic materials at cryogenic temperatures. The IMPATT diode pallet, when mounted on a diamond heat sink, offers the highest thermal conductivity and maximum thermal resistance even at room temperature (300 K), which will be significantly improved at cryogenic operations.

6.6 IMPROVEMENT OF THERMAL AND MECHANICAL PROPERTIES OF POTENTIAL MICROWAVE SUBSTRATE MATERIALS

6.6.1 Improvement in Thermal Characteristics

The design of current and future high-power microwave and electronic systems requires electronic substrates with high reliability, high thermal conductivity, and high mechanical integrity. High-power rf terminations, attenuators, switches, couplers, filters, and other microwave components require materials of high thermal conductivities to eliminate thermal stresses, maintain high reliability, and provide rapid heat transfer. But high thermal conductivity is possible only at cryogenic temperatures. Calculated values of thermal conductivity of potential electronic materials as a function of cryogenic temperatures

TABLE 6.4 Reliability (MTBF) for Diodes at Cryogenic Temperatures (h)

Temperature (K)	Schottky Barrier	Si-IMPATT	GaAs Varactor
300	1.4×10^{10}	4.7×10^{14}	9.8×10^{10}
200	7.2×10^{19}	5.8×10^{26}	3.5×10^{19}

TABLE 6.5 Improvement in Thermal Conductivity of Cryogenically Cooled Electronic Substrates (W/cm deg^{-1})

Temperature (K)	BeO	Aid	SiC	Al$_2$O$_3$
273	3.06	1.82	0.85	0.28
163	4.55	2.50	1.08	0.45
77	6.04	3.02	1.52	0.53
4.2	8.02	3.80	2.25	0.69

are shown in Table 6.5. In beryllium oxide (BeO) substrate both the thermal conductivity and thermal shock resistance are improved, but this hazardous material requires special handling and disposal techniques. Aluminium nitride (AlN) offers improved thermal performance under cryogenic operations and is a closer thermal expansion match to silicon, which provides higher reliability and structural integrity in thermally and mechanically stressed applications.

6.6.2 Structural Integrity of Cryogenically Cooled Dielectric Substrates

The advent of rocketry, nuclear medicines, space sensors, satellite communication, and supercomputers has intensified cryogenic research. Helium, deuterium, nitrogen, fluorine, oxygen, and methane are called cryogens when they are in a liquid state. These cryogens play key roles in cryogenically cooled components and systems. Conventional plastic substrates literally crumble at cryogenic temperatures. Only few substrates, such as alumina, Teflon or FFE with polymer resins known as FEP resins, are not adversely affected at cryogenic temperatures. In fact, both the electrical and mechanical properties of these resins improve as the operating temperature is reduced. Printed-circuit boards (PCBs) and cables insulted with Teflon have demonstrated superior mechanical properties at cryogenic temperatures.

Significant improvement in tensile strength, compressive strength, and flexural strength of unfilled FEP and FEP filled with 25% glass fiber can be verified from the data summarized in Table 6.6. The data are based on the

TABLE 6.6 Improvement in Mechanical Properties for Cryogenically Cooled Teflon Based Unfilled Substrate

Temperature (K)	Tensile Stress (psi)	Flexural Stress (psi)	Compressive Stress (psi)
300	4,000	2,900	1,500
200	6,500	9,600	13,300
90	17,850	24,700	28,800
77	18,000	26,700	30,500
20	23,750	36,300	35,900

work done by Du Pont scientists about 25 years ago. They indicate that the mechanical properties of even unfilled Teflon-based substrates significantly improve at cryogenic temperatures. Impact resistance of the parts molded from Teflon fluorocarbons remains impressive even at cryogenic temperatures. A typical impact resistance of 1.91 ft-lb/inch at 300 K is reduced to only 1.45 ft-lb/inch at an operating temperature of 20 K. Cryogenic research on dielectric substrates reveals that all fluorocarbons exhibit a remarkable flexural modulus of 680,000 psi at 77 K compared to that of 170,000 at 300 K.

6.7 HETEROJUNCTION BIPOLAR TRANSISTORS (HBTs)

Recent research and development activities on high-power HBTs fabricated on silicon substrates indicate that these devices, when operated at cryogenic temperatures, will exhibit significantly improved power performance and device reliability. Such cryogenically cooled devices offer great promise for high-speed microwave and MM-wave components and optoelectronic integrated circuits (ICs).

6.7.1 Performance Improvement of Cryogenically Cooled GaAs HBTs

The latest research indicates that a 3-terminal GaAs HBT device is best suited for high-power rf amplifiers operating up to 35 GHz. The power-handling capability of a silicon substrate is about 3.5 and 2.7 times higher than that of GaAs and InP substrate, respectively, even at room temperature (300 K). These power ratings can be further improved by a factor of at least 3:1 at 20-K operating temperature. The room temperature thermal conductivity of silicon is about 3 times greater than that of GaAs, but the electron mobility of GaAs is 6 times as large as that of silicon. This means that an HBT fabricated on a GaAs-on-Si composite structure will offer distinct advantages in terms of power output, gain, noise figure, and power-added efficiency at cryogenic temperatures. The latest research on solid-state devices indicates that an HBT on composite substrate is capable of delivering cw power levels as high as 6 W/mm emitter length. Such devices have great potential in the design of cryogenically cooled, high-power, MM-wave amplifiers.

6.7.2 Cryogenically Cooled HBTs for Optoelectronic Applications

Narrow-band HBT (NBHBT) devices using AlGaAs/GaAs with a base thickness of about 50 Å exhibit maximum small signal, common-emitter current gains (H_{fe}) in excess of 1400 at 300 K, 3000 at 77 K, and 3800 at 20 K approximately. Low-temperature measured data [1] indicate that the maximum current gain increases exponentially with decreasing temperature. Conventional HBTs have base thicknesses in the range of 500 to 1000 Å compared to a 50- to 200-Å range for NBHBT devices. In the case of GaAs FETs both the cw power and the pulsed power output levels are limited by the maximum drain current, while the power-handling capability of an NBHBT is strictly limited by the thermal effects. Thermal effects are significantly reduced at cryogenic temperatures. Cryogenic operations yield higher long-term reliability, regardless of whether the device is operating as a cw or pulsed signal. For a high-speed AlGaAs/GaAs-NBHBT with fewer emitter fingers, the maximum power density will be higher than 10 and 15 mW/nm^2 at operating temperatures of 200 and 100 K, respectively.

6.7.3 Performance of a Cryogenically Cooled Pseudo-HBT

The miniaturization of the silicon homogenous bipolar transistor has been limited by the tradeoff between emitter efficiency and base resistance. This limitation has been overcome by research on silicon-etched heterojunction bipolar transistors (HBTs) using microcrystalline Si or SiGe material and operating at cryogenic temperatures. Cryogenically cooled conventional HBTs offer higher injection efficiency, lower base resistance, and reduced emitter–base leakage current.

The new silicon-based homogeneous base transistor (HOB) structure is most ideal for low-temperature operations well below 40 K. The HOB uses a moderately doped emitter to avoid current freeze-out in a heavily doped bandgap–narrow-base structure.

These design aspects provide all the advantages of the conventional HBT. However, a pseudo-HOB does not require low-temperature processing, and, thus, can be fabricated with conventional processing techniques, including ion implantation and thermal annealing. The most distinct aspect of the pseudo-HOB is the negative exponential temperature dependence of the current gain (H_{fe}), which is a function of temperature-dependent effective bandgap voltage (V_{bg}). The effective bandgap voltage in the base resistance of the device is about 0.68 V at 300 K, 0.98 V at 77 K, 1.01 V at 50 K, and 1.04 V at 20 K. Published data indicate that this device with a self-aligned structure is capable of yielding 2500 times higher collector current at 77 K than a conventional HBT. This makes the cryogenically cooled pseudo-HOB devices most suitable for applications for bipolar/BiCMOS circuits.

6.7.4 Cryogenically Cooled Dielectric-Base HBT (DB-HBT)

Three terminal high-T_c superconducting transistors—including MESFETs, superconducting-base transistors known as flux flow transistors (FETs), and DB-HBTs—offer higher gain, improved isolation, higher speed, and lower power consumption than conventional solid-state devices. However, a DB-HBT (Figure 6.2) when operated at cryogenic temperatures offers significantly higher gain and isolation with operating voltage as low as 10 mV. Thus, when cooled to 77 K or lower, this device requires minimum operating voltage and

6.7 HETEROJUNCTION BIPOLAR TRANSISTORS (HBTs)

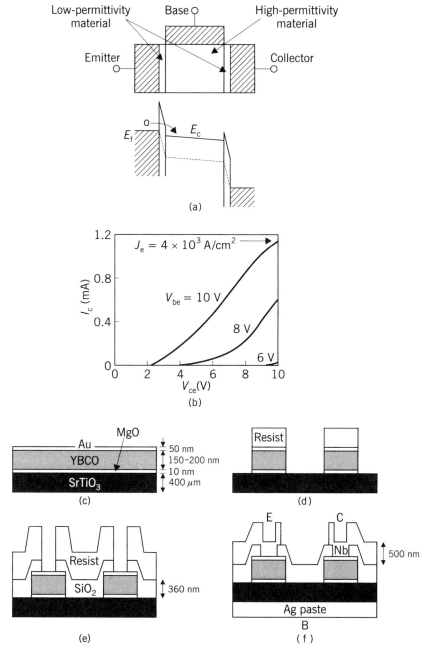

FIGURE 6.2 I-V characteristics and fabrication details of a 3-terminal superconducting dielectric base transistor [1]: (a) grounded emitter transistor configuration; (b) I_c-V_{ce} characteristics of a grounded emitter transistor; (c) emitter–collector electrodes; (d) etching process for electrodes; (e) deposition of SiO_2 insulating layer; (f) deposition and pattern of the Nb wire.

consumes less than 1 mA in the ON state, resulting in a power dissipation of less than 50 µW.

Structural details shown in Figure 6.2 reveal that the device employs ABC/NdGaO$_3$/SrTiO$_3$ heterostructures as the emitter and the collector junctions. The NdGaO$_3$ layer acts as the low-permittivity barrier and the SrTiO$_3$ acts as the superconducting-dielectric base. Potential high-T_c substrates that also include LaAlO$_3$, CeO$_2$, and MgO are the most ideal materials. These barriers can be epitaxially grown on the strontium titanate (SrTiO$_3$) substrate, and the ABC films can be grown on them with minimum cost and complexity. These superconducting substrates remain stable over the operating temperature range of 300–10 K. The transistor base current is less than 10 pA and the cryogenically cooled device offers a current gain of more than 200 dB. The DB-HBT offers transconductance greater than 50 mS/mm per emitter perimeter, higher current density at low operating voltages, and reduced Turn-on voltage all at 100-K operating temperature. This particular device is best suited where minimum power consumption over extended periods is the principal requirement.

6.8 TRANSISTORS FOR MICROWAVE/MM-WAVE APPLICATIONS

This section focuses on cryogenically cooled solid-state devices, including metal-semiconductor FETs (MESFETs), modulation-doped FETs (MOD FETs), high electron mobility transistors (HEMTs), and pseudomorphic-HEMTs (p-HEMTs), for possible use in microwave and MM-wave circuits and components. When operated at cryogenic temperatures these devices exhibit significantly improved gain, noise figure, and power-handling capability.

6.8.1 MESFET Devices

MESFETs are widely used in amplifiers and oscillators operating at microwave and MM-wave frequencies. Significant improvement in extrinsic transit frequency (F_t), cutoff frequency (F_{max}), and dc transconductance (G_m) is possible at cryogenic operations. Improvement in these parameters will lead to higher device gain (G), lower noise figure (NF), and higher power-added efficiency (η_{pae}).

The parameters shown in Figure 6.3 are defined as follows:

$$F_t = \frac{G_m}{2\pi(C_{gs} + C_{gd})} \quad (6.11)$$

$$F_{max} = \frac{F_t}{2}\frac{\sqrt{R_d}}{R_{in}} \quad (6.12)$$

6.8 TRANSISTORS FOR MICROWAVE/MM-WAVE APPLICATIONS

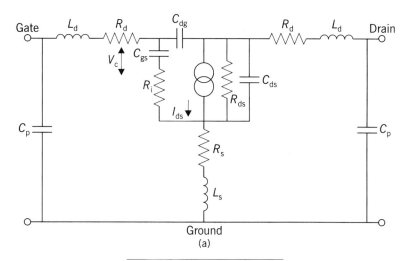

Element	Value	Units
L_g	0.01	nH
R_g	2.0	Ω
C_{gs}	0.25 (0.22)	pF
R_i	10.0	Ω
C_{dg}	0.027	pF
G_m	49 (41)	mS
R_{ds}	300 (330)	Ω
C_{ds}	0.05	pF
R_d	2.5	Ω
L_d	0.02	nH
R_s	2.5	Ω
L_s	0.05	nH
C_p	0.04	pF
τ	1.5	pS

Note: Values are at $V_{ds} = 3$ V, $I_{ds} = 30$ mA ($I_{ds} = 10$ mA). $I_{ds} = g_m V_{ce}^{-j\omega\tau}$.

(b)

Temperature (K)	Transconductance, G_m (mS)	Series Resistance (Ω)
300	48	2.25
200	57	1.41
100	60	0.51
77	62	0.35

(c)

FIGURE 6.3 (a) Equivalent circuit of MESFET. (b) Typical MESFET parameters (Plessey P-35-1140 MESFET). (c) Calculated values of transconductance and series resistance as a function of cryogenic temperature.

$$G = \left(\frac{F_{max}}{F}\right)^2 \quad (6.13)$$

$$\eta_{pae} = \left(1 - \frac{1}{G}\right)\eta_d \quad (6.14)$$

$$F_{min} = 1 + K2\pi FC_{gs}\frac{\sqrt{R_g + R}}{G_m} \quad (6.15)$$

where C_{gs} = gate–source capacitance
 C_{gd} = gate–drain capacitance
 K = constant, generally less than unity
 R_g = gate resistance
 R = source resistance
 R_d = drain resistance
 R_{in} = input resistance
 F_{min} = minimum noise figure of the amplifier
 G = power gain of the amplifer
 η_d = drain efficiency

The dc transconductance parameter G_m plays key role in determining the F_{max} and F_t parameters. However, significant improvement in the G_m and, thus, in the values of F_{max} and F_t is observed at cryogenic temperatures. Only slight improvement is seen in R_g, R_d, and R compared to that of G_m at cryogenic operations. Therefore, the performance improvement of a cryogenically cooled MESFET amplifier strictly depends on the higher values of transconductance.

Calculated values of G_m as a function of operating temperature for four distinct MESFET devices that can be used in the design of MESFET amplifiers are shown in Table 6.7. In all cases, the dc transconductance is improved when the operating temperature is reduced from 300 to 4.2 K. However, maximum improvement is possible between 77 and 4.2 K.

TABLE 6.7 Transconductance of MESFET Devices as a Function of Temperature (mS)

MESFET Device	300 K	77 K	4.2 K
A	25.3	30.3	41.5
B	23.4	27.6	39.2
C	27.2	36.3	44.2
D	31.1	38.4	45.8

6.8.1.1 Impact on Microwave and MM-Wave Amplifiers
Because transconductance increases at cryogenic temperatures, the parametric values of F_{max} and F_t are significantly improved. An increase in these parameters enhances the noise figure, power gain, and power-added efficiency of the MESFET amplifier regardless of the operating frequency (F). Improvement in gain and power-added efficiency as a function of operating temperature is evident from the data summarized in Table 6.8. One can see significant improvement in unilateral power gain and power-added efficiency at superconducting temperatures, assuming a drain efficiency of 40%. A higher drain efficiency will yield higher power-added efficiency.

6.8.1.2 Cryogenically Cooled MESFET Devices for High-Speed Integrated Circuits (ICs)
The small-signal saturation-region output conductance (g_{ds}) of a conventional three-terminal GaAs MESFET device is strictly dependent on both the operating frequency and temperature. The small-signal voltage gain in analog circuits and the propagation delay in digital circuits are dependent on this parameter. Calculated values of low-frequency output transconductance using the basic Curtice model [2] are shown in Figure 6.4 at various operating temperatures. The magnitude of this parameter is proportional to saturation electron velocity (v_s), peak electron velocity (v_p), and electron mobility (μ). All three parameters vary approximately inversely with the operating temperature. In GaAs devices the temperature-dependent time constant for emission of electrons for a midgap trap is shorter at higher operating temperatures.

The Curtice model defines the drain–source current (I_{ds}) in terms of transconductance (β), gate–source voltage (V_{gs}), threshold voltage (V_t), modulation (λ), drain–source voltage (V_{ds}), and saturation voltage (α). Typical values of transconductance and threshold voltage [3] as a function of temperature are shown in Figure 6.5, and similar values of λ and α at various cryogenic temperatures are shown in Figure 6.6.

TABLE 6.8 Improvement in Gain and Power-Added Efficiency in a MESFET Amplifier as a Function of Operating Temperature

Temperature (K)	Gain (dB)		Power-Added Efficiency (%)	
	35 GHz	20 GHz	35 GHz	20 GHz
300	11.06	15.92	28.82	38.96
200	12.29	17.15	30.26	39.20
77	13.20	18.06	31.24	39.38
20	14.29	19.88	33.12	39.98

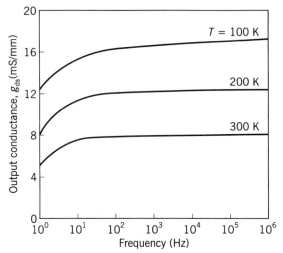

FIGURE 6.4 Projected low-frequency behavior of the output conductance for a MESFET device as a function of temperature, T [2].

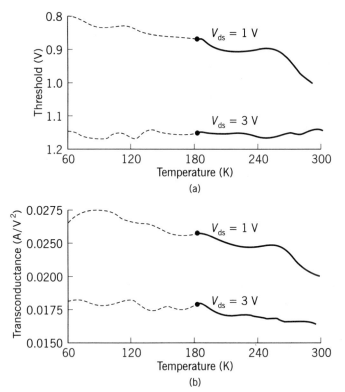

FIGURE 6.5 (a) Magnitude of threshold parameter (V_{TO}) as a function of temperature and bias voltage; (b) magnitude of transconductance parameter (β) as a function of operating temperature [3]. ---, projected values; —, computed values.

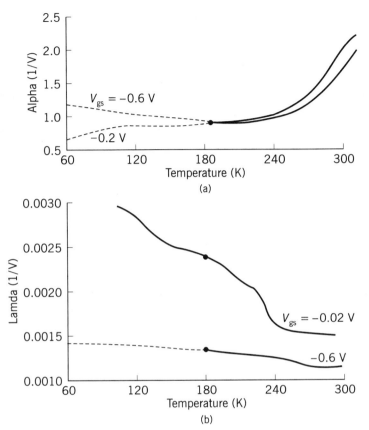

FIGURE 6.6 Impact of temperature and bias on (a) alpha and (b) lambda parameters [3]: ---, projected values; —, computed values.

6.8.2 Modulation-Doped FETs (MODFETs) and High Electron Mobility Transistors (HEMTs)

Low-frequency noise is a major drawback of a GaAs MESFET, because it limits the wideband operation. This problem can be overcome with cryogenically cooled two-dimensional electron gas FET (TEGFET) devices, which are also known as HEMTs or MODFETs.

6.8.2.1 Performance of Cryogenically Cooled TEGFETs
A TEGFET device offers improved performance over a MESFET at cryogenic temperatures, because of significantly increased electron mobility. However, its transconductance is lower than that of a pseudomorphic-HEMT (p-HEMT) even at cryogenic temperatures and, therefore, the TEGFET is not recommended for MM-wave components.

6.8.2.2 Cryogenically Cooled Pseudomorphic-MODFET (p-MODFET)

MODFET devices have demonstrated superior electron transport properties parallel to the heterointerface at superconducting temperatures. But the new MODFET, the so-called p-MODFET or strained-quantum-well (SQW)-MODFET device using InGaAs/AlGaAs structure, offers exceptionally high electron mobility and saturation velocity even at room temperatures (300 K), which are further improved at cryogenic temperatures.

Transconductance values as high as 310 and 380 mS/mm and drain currents as large as 290 and 310 mA have been realized at 300 and 77 K, respectively, in a InGaAs/AlGaAs-MODFET or p-MODFET device. This device is roughly 100% better than the similar GaAs/AlGaAs-MODFET in terms of critical performance parameters at cryogenic temperatures. The improvements in critical device parameters are due to higher electron velocity in the InGaAs semiconductor compound as compared to GaAs at superconducting temperatures. Electron mobility in the InGaAs/InAlAs-SQW-MODFET is even higher than that in InGaAs/AlGaAs-SQW-MODFET. Electron mobility as a function of cryogenic temperatures is shown in Figure 6.7. Specific dimensional details on InAlAs layer, undoped InAlGa spacer, InGaAs-SQW channel, InGAAs barrier, and the main substrate are

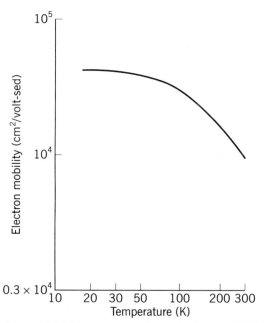

FIGURE 6.7 Electron mobility improvement in an InGaAs/InAlAs SQW-MODFET device as a function of temperature. Electron mobility increases with increase in channel thickness, which varies from 100 to 400 Å. Electron mobility also increases as the temperature is reduced.

not shown in Figure 6.7. Both the Hall mobility and carrier concentration increase as the operating temperature is reduced.

Impact of channel temperature on Hall mobility and carrier concentration is evident from the data summarized in Table 6.9. Electron mobility increases with the increase in SQW channel thickness and with the decrease in operating temperature. The Hall mobility and carrier concentration of the SQW-MODFET are similar to those of a conventional MISFET (metal-insulator semiconductor FET) device at cryogenic temperatures.

6.8.2.3 Overall Impact of Indium Mole Fraction and Operating Temperature on Dc Transconductance

Computer modeling and device research activities indicate that dc transconductance improves with the increase in In-mole fraction and with the decrease in operating temperature. Peak dc transconductance [4] is about 360 mS/mm at a gate voltage of +0.2 V and at an operating temperature of 15 K. The impact of In mole fraction (x) in the $In_xGa_{1-x}As/Al_xGa_{1-x}As$-MODFET structure and operating temperature is evident from the data summarized in Table 6.10. For a mole fraction greater than 20%, the increase in transconductance is very small. However, significant increase is realized when the operating temperature is reduced from 300 to 77 K. Maximum transconductance values can be expected only at lower cryogenic temperatures.

TABLE 6.9 Impact of Channel Temperature on Hall Mobility and Carrier Concentration

Temperature (K)	Hall Mobility ($cm^2/V \cdot s$)	Carrier Concentration (cm^{-2})
300	5,000	1.3×10^{12}
77	100,000	5.2×10^{11}
10	210,000	5.2×10^{11}
4.2	400,000	5.2×10^{11}

TABLE 6.10 Impact of Indium Mole Fraction and Operating Temperature on Device Transconductance (mS/mm) [4]

In-mole (x)%	300 K	77 K
5	253	303
10	234*	276
15	270	360
20	310	380

* Indicates optimum reduction in transconductance value most likely due to underetching of the gate recess during the etching process.

The InGaAs/AlGaAs-SQW-p-MODFET exhibits superior dc and rf parameters over the conventional GaAs/AlGaAs-MODFET at cryogenic temperatures. Improved performance is possible even with low In mole fraction, but with better carrier confinement in the InGaAs quantum-well layer. Furthermore, low-frequency ($1/f$) noise performance is drastically improved at cryogenic operations. Outstanding performance and cryogenic stability make this device very attractive for high-speed logic and low-noise rf applications.

6.9 HEMT AND p-HEMT DEVICES

High electron mobility transistor (HEMT) and pseudomorphic-HEMT (p-HEMT) devices (Figure 6.8) when operated at cryogenic temperatures offer significant improvement in system performance, particularly at MM-wave frequencies. Improvement in unity current gain frequency (f_T), maximum frequency of oscillation (f_{MAX}), gain (G), and noise figure (NF) have been observed in these device at MM-wave frequencies.

6.9.1 Unique Capabilities

Improved dc transconductance, unity current gain frequency, and maximum frequency of oscillation are observed even at room temperature compared to MODFETs devices. As the operating temperature is reduced, significant improvement is seen in these parameters. A standard HEMT structure consists of doped GaAs, AlGaAs, undoped AlGaAs, and undoped GaAs layers on an undoped substrate. However, an inverted-HEMT (i-HEMT) comprising of a InAlAs-inserted channel in the InAlAs/InGaAs structure offers superior MM-wave performance in terms of extrinsic transconductance and electron mobility over the standard HEMT device at cryogenic temperatures [5]. Structural details of the i-HEMT device are shown in Figure 6.9. Insertion of InGa layer improves the electron transport properties significantly, particularly, at cryogenic temperatures. The kink effect occurs only at 4.2 K and at low drain–source voltage. This effect increases slightly the gate current, but does not affect other performance parameters.

Another version of the HEMT device is the AlGaN/GaN-doped channel HEMT (DC-HEMT) using gallium nitride (GaN) doped channel. This heterostructure field effect transistor (FET) with 1-μm gate length as depicted in Figure 6.10 exhibits reduced parasitic series resistance. The DC-HEMT offers microwave performance comparable to that of the state-of-the-art GaAs MESFET at cryogenic temperatures. This device offers extrinsic transconductance greater than 120 mS/mm and improved reliability even at maximum drain–source saturation current (I_{DSS}) exceeding 600 mA/mm under cryogenic environments. The large effective mass of electrons in GaN makes the electron

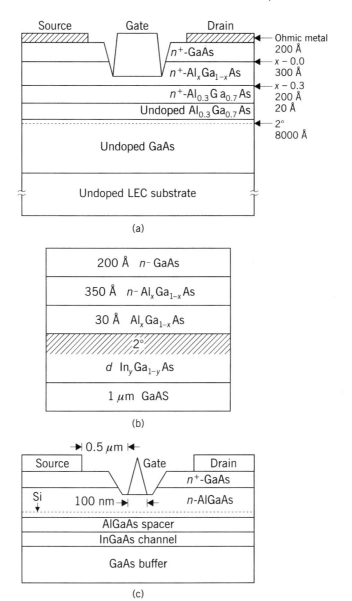

FIGURE 6.8 Cross-sectional view of (a) standard HEMT; (b) submicron p-HEMT; (c) planar doped p-HEMT.

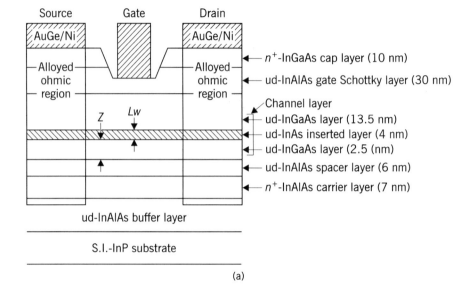

FIGURE 6.9 (a) Structural details of InAs inserted channel inverted-HEMT (i-HEMT) device; (b) extrinsic transconductance and electron mobility of i-HEMT as a function of cryogenic temperature.

mobility less sensitive to the channel doping spread [6]. Both electron (μ_e) and hole mobility (μ_h) improve as the operating temperature is reduced.

Planar doped p-HEMT or standard p-HEMT with AlGaAs/GaAs heterostructure offers superior performance over the standard HEMT at cryogenic temperatures. However, at cryogenic temperatures AlGaAs/GaAs HEMT suffers from collapse of the drain I-V characteristics, small conduction band discontinuity, and threshold voltage shift caused by deep trap levels in the n-AlGaAs layer. These problems can be minimized in AlGaAs/InGaAs HEMT devices due to larger conduction band discontinuity without any degradation of two-dimensional electron gas (2-DEG) concentration. The high electron density of AlGaAs/InGaAs HEMT is particularly attractive for achieving good performance at MM-wave frequencies, and is further improved at cryogenic temperatures.

A p-HEMT with T-shaped gate configuration (Figure 6.11) offers high-power added-efficiency, improved rf power output, and ultralow noise perfor-

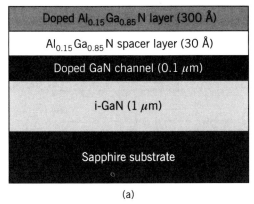

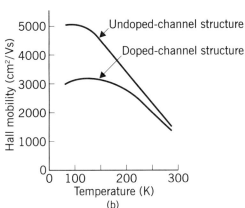

FIGURE 6.10 (a) Structural details of a cryogenically cooled AlGaN/GaN-doped-channel HEMT; (b) impact of cryogenic temperature on Hall mobility in an HEMT device. The AlGaN/GaN material system is less sensitive to channel doping.

mance when cooled to cryogenic temperatures. The noise performance of this device at 20 K is superior to that of AlGaAs/GaAs-HEMT under cryogenic operations.

SUMMARY

The performance capabilities of Si and GaAs varactor diodes, PIN diodes, Schottky barrier diodes, HBTs, MESFETs, MODFETs, HEMTs, i-HEMTs, DC-HEMTs, p-HEMTs, and SQW-MODFETs can be significantly improved at cryogenic temperatures. In addition to rf improvement, cryogenic operation improves reliability, thermal efficiency, and operational longevity of the device under high-power conditions.

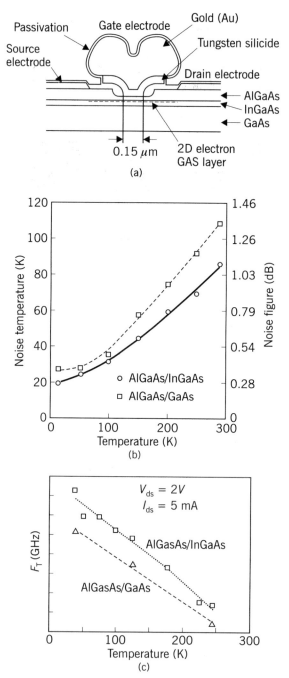

FIGURE 6.11 RF performance of cryogenically cooled p-HEMT with T-shaped gate structure: (a) structural details; (b) noise performance; (c) unity current gain parameter (F_T) as a function of operating temperature. $F_T = g_m/(6.28)(C_{gs} + C_{gd})$, where g_m is transconductance, C_{gs} is gate–source capacitance, and C_{gd} is gate–drain capacitance.

REFERENCES

1. K. I. Anastaslou et al. Low temperature characteristics of high current-gain graded-emitter AlGaAs narrow-band HBT. *IEEE Electron Device Lett.* 13:414–417 (1992).
2. W. R. Curtice. A MESFET model for use in the design of GaAs-ICs. *IEEE Trans. MTT* 28:448–456 (1980).
3. J. Rodrigue–Tellez et al. Simulation of temperature and bias dependencies of β and V_{TO} of GaAs MESFET devices. *IEEE Trans. Electron Devices* 40:1730–1734 (1993).
4. A. A. Ketterson et al. Characterization of InGaAs/AlGaAs-pseudomorphic modulation-doped field-effect transistor. *IEEE Trans. Electron Devices* ED-33:564–569 (1986).
5. A. Tatushi et al. Kink effect in an InAs-inserted channel, InAlAs/InGaAs-inverted HEMT at low temperatures. *IEEE Electron Device Lett.* 17:378–382 (1996).
6. M. A. Khan et al. Microwave operation of GaN/AlGaN-doped channel heterostructure field-effect transistors. *IEEE Electron Device Lett.* 17:325–326 (1996).

CHAPTER SEVEN

Application of Superconductor Technology to Components Used in Radar, Communication, Space, and Electronic Warfare

This chapter describes the performance of cryogenically cooled rf components for possible applications to radar, communication, space, and electronic warfare (EW) systems. Components that can effectively use superconductor technology include amplifiers, antennas, filters, mixers, multipliers, oscillators, receivers, and high-power transmitters. Particular emphasis is placed on FM phase noise and noise figure (NF). Improvement in noise figure for microwave components as a function of cryogenic temperatures is clearly illustrated in Figure 7.1.

7.1 CRYOGENICALLY COOLED SOLID-STATE AMPLIFIERS

GaAs MESFET, HBT, HEMT, and p-HEMT amplifiers are widely used for radar, communication, space, and EW applications. Significant improvement in gain, noise figure, and power consumption have been observed at cryogenic temperatures. Performance capabilities of potential cryogenically cooled rf amplifiers and microwave amplifiers using both the MMIC and HTSC technologies are described. Emphasis is on maximum available gain and noise figure of the microwave and MM-wave amplifiers. Improvement in performance is strictly due to an increase in electron mobility (μ) and transconductance (g_m) of the solid-state devices with a decrease in operating temperature. An increase in these parameters will lead to higher gain and lower noise figure in the cryogenically cooled solid-state rf amplifiers. Gain of an X-band, 3-stage,

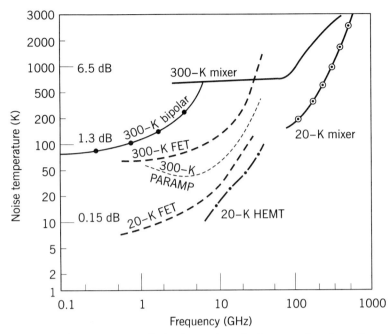

FIGURE 7.1 Noise temperature limits for various devices as a function of cryogenic temperature and operating frequency.

GaAs MESFET amplifier as a function of bias level and operating temperature is illustrated in Figure 7.2.

7.1.1 Performance of Superconducting GaAs Low-Noise Amplifiers

Commerically available microwave amplifiers using submicrometer GaAs MESFET devices offer high gain and low noise figure, but over a relatively narrow band. Integration of superconductor technology can improve both the gain and noise figure over a wide frequency spectrum with minimum power consumption. Performance capabilities of wide-band and narrow-band microwave amplifiers as function of cryogenic temperature are shown in Table 7.1.

7.1.2 HTSC Heterojunction Bipolar Transistor (HBT) Power Amplifiers

In general, GaAs HBT amplifiers exhibit higher power gain than Si bipolar amplifiers orGaAs MESFET amplifiers at microwave frequencies. The use of a heterojunction to support the hole injection into the emitter together with the superior electronic properties of GaAs compound results in greatly improved power gain at microwave frequencies. Since, the maximum current in a well-designed GaAs HBT amplifier is greater than GaAs MESFET amplifier, values

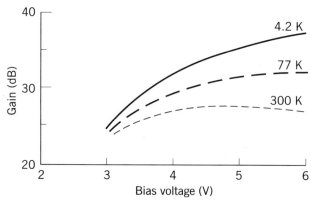

FIGURE 7.2 X-band 3-stage GaAs MESFET amplifier gain as a function of bias voltage and temperature.

of both the power output and the power density are higher in GaAs HBT amplifier. These parameters are further improved at cryogenic operating temperatures, as illustrated in Table 7.2.

The circuit model for a common-emitter HBT amplifier is shown in Figure 7.3. The maximum available gain (MAG), maximum stable gain (MSG), and power density (Pd) all increase as the operating temperature is reduced. Both the maximum power output and power density are determined by the junction temperature. However, HBT amplifier performance is limited by the thermal effects long before the electronics limitations are reached. But the thermal limits are significantly improved at 77 K compared to 300 K. The cryogenically cooled GaAs HBT amplifiers are capable of yielding room-temperature cw

TABLE 7.1 Performance of Cryogenically Cooled Amplifiers

Temperature (K)	Typical Gain (dB)	Typical Noise Figure (dB)
Wideband Amplifier (8–18 GHz)		
300	22	2.08
200	24	1.59
77	27	0.87
10	28	0.75
Narrowband Amplifier (9.5–10.5 GHz)		
300	22	1.25
200	25	0.99
77	29	0.35
10	31	0.26

TABLE 7.2 Performance of X-Band, GaAs HBT at Cryogenic Temperatures

Performance Parameters	Temperature	
	300 K	77 K
Gain (dB)	8	12
CW power density (W/mm)	4	8
Power output (watt)	0.8	1.4

power density and pulsed power density in excess of 4 and 10 W/mm, respectively, at 10 GHz frequency. The power output levels are improved at least by a factor of 2:1 at 77 K (Table 7.3). Both the power output and the power-added efficiency improve at 77-K operating temperature. As the MESFET device area increases, the amplifier exhibits even better thermal performance at superconducting temperatures.

The cw operation of an HBT amplifier is dependent on the number of emitter figures, the spacing between the fingers, and the operating temperature. Optimum thermal isolation is possible with 20 emitter fingers having a spacing of 35 μm. The same optimum thermal isolation is possible with spacing as small as 20 μm at 77 K. This means that the higher thermal isolation needed for improved reliability is possible only at cryogenic temperatures. Preliminary calculations indicate that an MMIC amplifier using GaAs HBT devices with dimensions of 1.7 mm × 1.9 mm can deliver cw power output exceeding 2.5 W at 300 K and 4.5 W at 77 K with corresponding power-added efficiencies of 40 and 52%, respectively. Cryogenically cooled GaAs HBT amplifiers are more

TABLE 7.3 Performance of X-Band, HBT Amplifiers With Various Configurations

Parameters	Temperature	
	300 K	77 K
Common-Base Configuration, Single-Stage, 20 Emitter Fingers		
P_{out} (W)	1.43	2.85
Power-added eff. (%)	31	43
Common-Emitter Configuration, Single-Stage, 20 Emitter Fingers		
P_{out} (W)	1.05	2.25
Power-added eff. (%)	43	50
Common-Emitter Configuration, Two-Stage, 60 Emitter Fingers		
P_{out} (W)	2.82	4.43
Power-added eff. (%)	45	51

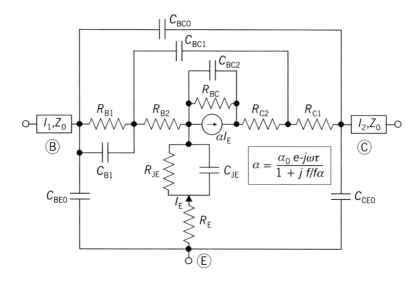

Element	Value	Element	Value
R_{B1}	5.47 Ω	C_{BC1}	0.22 pF
R_{B2}	1.90 Ω	C_{BC2}	0.02 pF
R_{C1}	0.19 Ω	C_{CE0}	0.08 pF
R_{C2}	0.78 Ω	C_{BE0}	0.04 pF
R_E	1.26 Ω	α_0	0.84
R_{JE}	0.65 Ω	f_α	31 GHz
R_{BC}	1.MΩ	τ	2.56 ps
C_{B1}	1.68 pF	l_1	0.01λ
C_{JE}	2.37 pF	l_2	0.01λ
C_{BC0}	0.07 pF	z_0	50 Ω

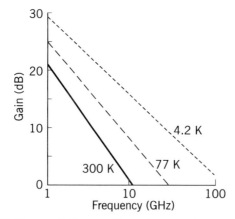

FIGURE 7.3 HBT monolithic amplifier gain at various temperatures: (a) equivalent circuit diagram, (b) typical parameters, and (c) gain as a function of frequency and temperature.

attractive for monolithic integration than GaAs MESFET amplifiers because of the highly reproducible nature of their vertical device category.

7.1.3 Noise Performance of GaAs HBT Amplifiers at Cryogenic Temperatures

The device dimensions, number of fingers, and spacing between them will be different for HBT amplifiers optimized for low noise and high gain. Both the noise temperature and gain will significantly improve as the operating temperature is reduced. Gain and noise temperature of a cryogenically cooled GaAs MMIC amplifier without feedback at 12 GHz and with feedback at 14 GHz as a function of operating temperatures are shown in Figure 7.4. Significant improvement, particularly in noise temperature is observed at low cryogenic temperatures. Cryogenically cooled MMIC amplifiers using MESFET devices

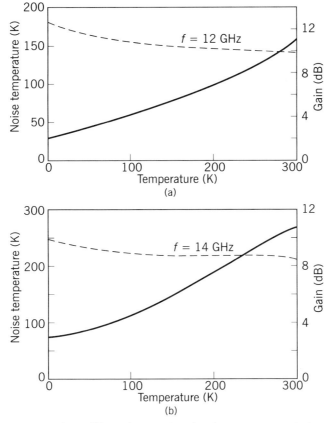

FIGURE 7.4 MMIC amplifier gain (– – –) and noise temperature (—) as a function of cryogenic temperature: (a) without feedback loop; (b) with feedback loop.

have potential applications in communication systems, MM-wave missile seekers, and space imaging sensors, because cryogenic operations not only improve rf performance and reliability, but also reduce power consumption, weight, and size.

7.1.4 Cryogenically Cooled HEMT, p-HEMT, and DH-HEMT Amplifiers

High electron mobility transistor (HEMT) amplifiers, p-HEMT amplifiers, and DH-HEMT amplifiers offer higher gain and lower noise figure than MESFET amplifiers at MM-wave frequencies. These amplifiers when operated at cryogenic temperatures are most suitable for MM-wave space radars, MM-wave missile seeker receivers, and space-imaging sensors. The higher gain and lower noise figures of these cryogenically cooled HEMT amplifiers offer substantial reduction in weight, size, and transmitter power requirements, which leads to considerable overall cost reduction.

Higher gain and lower noise figure are possible because of higher transconductance, faster response time, lower drain–source resistance, reduced gate resistance, lower source resistance, and reduced drain noise conductance at cryogenic temperatures. Performance improvement of a cryogenically cooled p-HEMT amplifier [1] is evident from the data summarized in Table 7.4. Performance improvement is possible in both a balanced p-HEMT amplifier and a p-HEMT amplifier with negative feedback at cryogenic operations, as illustrated in Figure 7.4. A balanced p-HEMT amplifier consumes more dc power and requires more discrete components, whereas the negative feedback loop tends to minimize the amplifier gain fluctuations caused by the variations in the device transconductance at cryogenic operations. However, these variations can be significantly reduced at lower cryogenic temperatures (less than 20 K).

Remarkable improvement in the associated gain and noise figure of a cryogenically cooled AlGaAs/GaAs-HEMT amplifier with T-gate structure is evident from Figure 7.5. This single-stage amplifier with T-gate dimensions of 1.1 µm × 300 µm offers a noise figure as low as 0.4 dB at 18 GHz and at a cryogenic temperature of 20 K, which is strictly due to higher device transconductance and cutoff frequency at cryogenic temperatures. One can also expect higher gain because of significant improvement in the transconductance and

TABLE 7.4 Performance Parameters of a Cryogenically Cooled p-HEMT Amplifier

Parameters	12 GHz		18 GHz	
	300 K	20 K	300 K	20 K
Gain improvement (dB)	0	2	0	1.2
Noise figure (dB)	0.86	0.29	1.3	0.40
Power consumption (W)	0.068	0.047	0.092	0.060

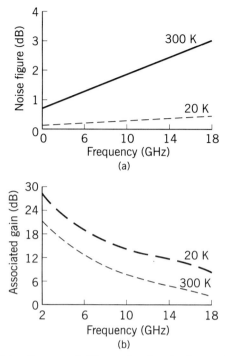

FIGURE 7.5 (a) Noise figure and (b) associated gain for an AlGaAs/GaAs HEMT amplifier as a function of temperature and frequency.

cutoff frequency at lower operating temperatures. Lower operating temperatures lead to a considerable reduction in drain resistance, which will further lower the noise figure.

Performance behavior of an HEMT or p-HEMT amplifier at cryogenic temperatures is very complex due to capture of the carriers in the device. However, replacement of the doped layers of n-GaAs and AlAs within the AlGaAs/GaAs-HEMT structure can overcome the collapsing problem. In a p-HEMT amplifier, a superlattice layer can easily overcome the collapsing problem, which is often seen at lower cryogenic temperatures in high-performance devices.

7.1.5 Cryogenically Cooled Operational Amplifiers

All electronic and rf systems suffer from noise problems, regardless of the operating frequency and temperature. In wideband systems designed to process very weak signals, noise problems are critical. Operational amplifiers (OPAMP) are widely used in integrated circuits (ICs) and can be treated as internal noise generating sources. The noise can be either a white noise or thermal noise known as Johnson noise or frequency-dependent shot noise.

150 SUPERCONDUCTOR TECHNOLOGY IN COMPONENTS

Thermal noise is the most dominant component of the overall noise, which can be reduced at cryogenic temperatures.

7.1.5.1 Noise Performance of Cryogenically Cooled OPAMP

The models of operational amplifiers shown in Figure 7.6 can be used to predict noise at cryogenic temperatures. Spectral density is widely used by IC manufacturers to specify the noise performance of their products. The noise spectral density is simply the root mean square (rms) value of the noise voltage (E_n) or noise current (I_n) expressed as a voltage or current per root hertz, namely, volts/$\sqrt{\text{hertz}}$ or ampere/$\sqrt{\text{hertz}}$. In an FET-OPAMP, bias current typically doubles for each 10-K increase of the FET junction temperature, which increases the current noise by a factor of $\sqrt{2}$ for each 10-K rise in junction temperature. The noise can be reduced by a factor of $1/\sqrt{2}$, if the junction temperature is reduced by 10 K. Bipolar OPAMPs exhibit higher current noise than FET OPAMPs, but are superior in their voltage noise. Thermal noise results from random motion of free electrons in a conductor caused by thermal agitation. The rms thermal noise in an OPAMP can be expressed as

$$E_n = \sqrt{4kTRB} \tag{7.1}$$

where k = Boltzmann constant
 T = absolute temperature (K)
 R = thermal resistance (Ω)
 B = system bandwidth (Hz)

Calculated values of rms thermal noise as a function of cryogenic temperature are shown in Table 7.5.

7.1.6 Cryogenically Cooled Parametric Amplifiers (PARAMP)

In a PARAMP, the nonlinear reactance is pumped at a frequency f_p and a signal is applied at a frequency of f_s. The dissipation at idler frequency f_i ($f_i = f_p - f_s$) produces negative resistance at the input of the amplifier. Negative resistance is used as a gain mechanism in the PARAMP. When the signal and pump frequencies are different, the PARAMP is called a nondegenerative amplifier. The performance of a PARAMP, which includes maximum exchangeable gain, noise figure, and conversion efficiency, is strictly dependent on the cutoff frequency (w_c) and series resistance (R_s) of a varactor diode. The series resistance includes the contributions from depletion layer thickness, contact terminal, and lead wire. The cutoff frequency f_c ($f_c = 1/2\pi R_s C_j$, where C_j is the diode junction capacitance) is inversely proportional to the series resistance, which decreases as the operating temperature is reduced. The perfor-

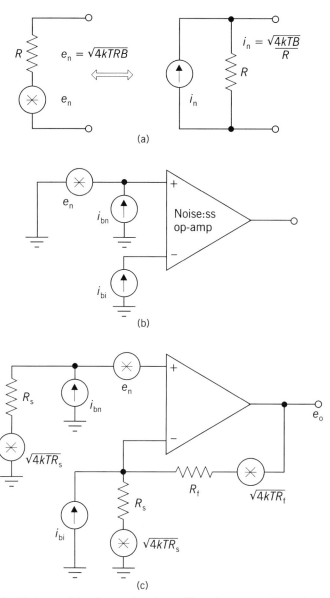

FIGURE 7.6 Noise models of operational amplifiers that can realize noise improvement at cryogenic temperatures: (a) interchangeable resistor; (b) typical parameters; (c) external resistor.

TABLE 7.5 RMS Thermal Noise of an OPAMP as a Function of Operating Temperature (nV/$\sqrt{Hz}$)

Temperature (K)	$R=1$ (Ω)	$R=0.001$ (Ω)
300	128.7	4.07
200	105.2	3.32
100	74.3	2.35
77	65.3	2.06
4.2	14.8	0.47

mance of a PARAMP is strictly a function of the cutoff frequency of the varactor used.

7.1.6.1 Performance Parameters The maximum exchangeable gain $(G_e)_{max}$ is the most important parameter of the PARAMP, and is defined as

$$(G_e)_{max} = \frac{m_1 w_c}{2 w_s} \qquad (7.2)$$

where m_1 = fundamental frequency modulation ratio. As stated earlier, the cutoff frequency improves as the operating temperature is reduced. Improvement in cutoff frequency and gain as a function of temperature is evident from the calculated values shown in Table 7.6. These calculations assume a room temperature cutoff frequency of 500 GHz, sinusoidal current pumping, abrupt junction GaAs varactor diode, and modulation ratio of 0.25. Sinusoidal current pumping [2] provides sinusoidal elastance and must be initiated at optimum pump frequency. Square waves offers the highest value of the modulation ratio, which requires the highest pump frequency and leads to high cost.

Another important performance parameter is the noise figure or noise temperature of the PARAMP. The noise temperature of a PARAMP is directly proportional to diode operating temperature. Other parameters, such as diode series resistance, signal frequency, cutoff frequency, minimum elastance, and modulation ratio, have a secondary impact on the noise temperature. Improvements in series resistance and cutoff frequency at cryogenic temperatures are primarily responsible for low noise temperature. Preliminary calcula-

TABLE 7.6 Maximum Exchangeable Gain as a Function of Temperature

Temperature (K)	Cutoff Frequency (GHz)	$(G_e)_{max}$ (dB)
300	500	7.9
200	800	10.1
100	1200	11.8
40	1500	12.7

tions indicate that the noise temperature is about 340 K at a diode temperature of 300 K, 128 K at a diode temperature of 77 K, and 50 K at a diode temperature of 40 K, when the signal frequency is equal to $0.4m_1w_c$.

7.2 ANTENNAS

Ideal antennas must have high radiation efficiency over a broad frequency spectrum with minimum weight, size, and cost. Wireless communication systems require highly efficient electrically small antennas to fit into compact housing. However, an electrically small antenna possesses a very low radiation resistance generally associated with a very high inductive reactance component. Because of excessive power dissipated in the radiating element and its matching network, the radiation efficiency of an electrically small antenna is extremely low. However, integration of superconducting technology into the radiating elements will lead to radiation efficiency as high as 75% at an operating temperature of 77 K. The efficiency can be further improved if the operating temperature is reduced to 10 or 20 K, but at the expense of much higher cost, weight, and power consumption. Application of superconducting technology to an electrically small, printed-circuit antenna employing self-resonant radiating elements offers the highest radiation efficiency because of the zero reactive component.

7.2.1 Cryogenically Cooled, Electrically Small, Printed-Circuit Antennas Using Meander-Line Radiating Elements

Higher radiation efficiency and wider operating bandwidth are possible with cryogenically cooled, electrically small, printed-circuit antennas using meander-line radiating elements. Structural details of such an antenna using superconducting technology and printed-circuit technology are shown in Figure 7.7. The currents in the parallel meander sections [3] are in phase, which leads to higher radiation efficiency. The meander-line radiating element uses a EuBaCuO (EBCO) superconducting thin film ($T_c = 87.8$ K), which is deposited on an MgO substrate ($e_r = 9.41$). The film can be patterned into a meander-line structure employing conventional photolithography technology. The microstrip transmission line (MTL) and the EBCO thin-film radiator are electromagnetically coupled as depicted in the Figure 7.7.

The radiation efficiency is dependent on the radiation resistance, matching network efficiency, coupling efficiency, input impedance of the antenna and operating temperature. The gain of the superconducting, high-efficiency, electrically small antenna (HEESA) can decrease due to weakness of the superconductivity in the film if proper thickness of the film is not selected. The gain of a superconducting transmit antenna will be lower than that of a receive antenna with the same physical dimensions, because of the difference in the

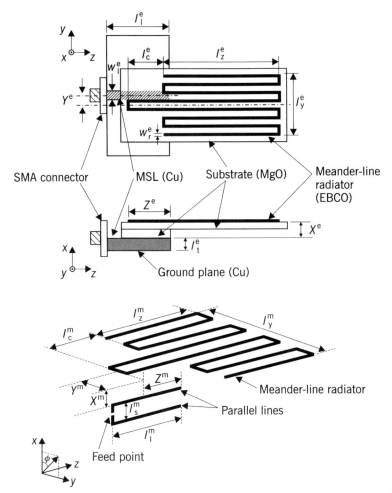

FIGURE 7.7 Cryogenically cooled, thin-film, electrically small antenna for wireless communication [3].

input power levels. However, the radiation efficiency (η_r) is dependent on radiation resistance and antenna loss resistance and is given as

$$\eta_r = \frac{R_r + R_a}{R_r} \tag{7.3}$$

where R_r = radiation resistance
R_a = antenna loss resistance

Radiation resistance is much greater than antenna loss resistance at superconducting temperatures, which leads to very high radiation efficiency at cryogenic temperatures. Typical performance parameters of a cryogenically cooled, electrically small, 936-MHz wireless antenna are summarized in Table 7.7. One can see significant performance improvement of the superconducting antenna over copper antenna at the same operating temperature of 77 K. It is evident from the performance comparison data that the 3-dB bandwidth is significantly reduced for both the transmit and receive antenna.

7.2.2 Cryogenically Cooled, Thick-Film, Printed-Circuit Antenna

Uncooled printed-circuit, electrically small antennas less than $\lambda/16$ in length suffer from low radiation efficiency, becasue of excessive losses in feed line, matching network, and radiating elements. Integration of superconducting technology in these antennas will significantly reduce the losses, leading to very high radiation efficiency. Radiation efficiency also depends on the type of feed and matching network and the quality of the thick film used. Studies performed on thick-film, printed-circuit antennas reveal that significant improvement in radiation efficiency of a low-profile, electrically small antenna is possible with high-quality, superconducting thick films.

7.2.2.1 Performance Capabilities of an Electrically Small, Printed Circuit Loop Antenna Using YBCO Thick Films
Significant improvement in the performance of electrically small, printed-circuit antennas using superconducting YBCO thick films has been observed in case of a YBCO thick-film loop antenna [4] and a thick-film dipole antenna [5]. As stated earlier, superconducting thick films of good quality offer superior performance over thin films. However, the patterned films with edge damage and current accumulation at the edges tend to show large performance degradation, particularly in thin films. The radiation efficiency of a superconducting antenna is proportional to radiation resistance, which is inversely proportional to the surface resistance. In superconducting antennas using YBCO thick films, the surface resistance is a function of operating temperature as well as of crossover

TABLE 7.7 Calculated Performance Data for a Cryogenically Cooled, Electrically Small, Wireless Antenna

Performance Parameters	Copper Antenna at 77 K	Superconducting Antenna at 77 K	
		Transmit	Receive
Radiation (η_r) %	6.2	19.22	74.35
3-dB bandwidth (%)	1.1	0.28	0.12
Gain (dB)	-11.1	-5.2	-0.82
Input power (dBm)	$+5$	$+5$	-27

frequency. Crossover frequency is defined as the frequency at which the surface resistance of the superconducting film becomes equal to that of copper film at 77 K. The surface resistance (R_s) of a superconducting YBCO film increases with frequency as f^2, while that of a normal copper increases as $\sqrt{f}$.

The radiation efficiencies of a copper film antenna on Duroid substrate (e_r), YBCO thick-film antenna on yttria-stabilized zirconia (YSZ) substrate ($e_r = 20$–30), and YBCO thick-film antenna on alumina substrate ($e_r = 9.8$) at cryogenic temperatures are shown in Figure 7.8. Specific details on the superconducting thick-film loop antenna of various configurations using specific substrates and appropriate matching circuits are also shown in Figure 7.8. Preliminary calculations indicate that the radiation efficiency of a copper loop antenna at 77 K is about 1.1% over VHF frequencies, but better than 10% over the 140- to 160-MHz range. However, the radiation efficiency of a superconducting thick-film loop antenna is greater than 70% at 77 K over the 400- to 500-MHz frequency range.

7.2.2.2 Performance Capabilities of Printed Circuit Dipole Antennas Using Thick Films of YBCO at 77 K

High antenna gain is possible with an array consisting of several closely spaced, printed-circuit dipoles alternately crossed by an integrated series-fed network [5]. To realize both high gain and radiation efficiency, dipole elements, feed network, and matching circuit must be made from thick films of YBCO superconducting material. The superconducting array offers higher gain than a cryogenically cooled copper or silver array, as illustrated in Figure 7.9. The radiation efficiency is dependent on element-to-element spacing, number of elements, radiator material, and operating temperature.

Calculated values of gain and efficiency for a silver array antenna and superconducting array antenna both operating at 10 GHz and at 77 K are summarized in Table 7.8. A 4-element, 1-GHz array using thick dipoles made from copper, silver, and YBCO thick films of same thickness demonstrated a gain of 1.90, 2.91, and 7.25 dB, respectively, at an operating temperature of 77 K. This clearly indicates that the superconducting array using thick films of YBCO yielded about 5.3 dB more gain over copper and about 4.3 dB more gain over silver at 77 K.

7.2.2.3 Potential Feed Networks for Superconducting Microstrip Antennas

Superconducting printed-circuit antennas using high-permittivity substrates offer high overall efficiency. But, high-permittivity substrates create feed problems because of extremely high patch edge impedance ranging from 500 to 1500 Ω, which leads to an impedance-matching problem. Superconducting microstrip printed-circuit antennas that are directly coupled and gap-coupled [6] to microstrip transmission lines offer a solution to this problem.

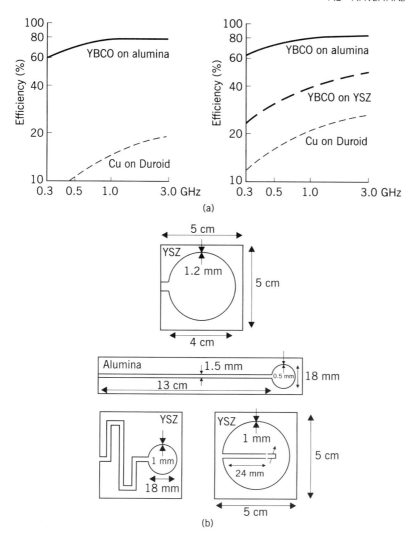

FIGURE 7.8 (a) Efficiency of electrically small, printed circuit antennas using thin films on various substrates at 77 K [4]. (b) Schematics of a loop antenna using YBCO superconducting thick films on various substrates [5].

The efficiency of a directly coupled antenna is illustrated by the feed VSWR. For a gold-plated patch antenna the VSWR is 1.39 at 77 K and 1.37 at 20 K, whereas the VSWR for a HTSC patch antenna is less than 1.17 at 77 K and less than 1.13 at 20 K over a 28- to 30-GHz frequency spectrum. Table 7.9 shows the gain capabilities of two distinct feed methods for cryogenically cooled microstrip patch antennas as a function of temperature. The gain of a printed-circuit microstrip antenna is the lowest at room temperature (300 K), regardless of the feed methods used.

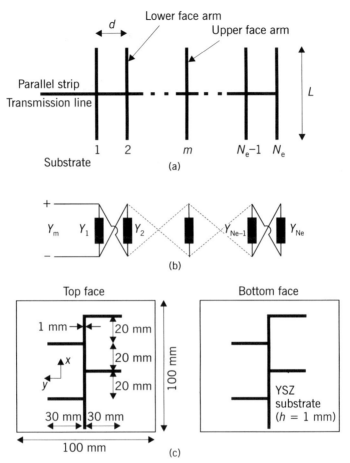

FIGURE 7.9 Cryogenically cooled, 1-GHz superconducting, series-fed, end-fire array using thick films [5]: (a) uniformly spaced cross dipoles; (b) equivalent circuit (Y = admittance, d = spacing, L = length, Ne = element number); (c) physical dimensions and layout.

TABLE 7.8 Relative Gain and Overall Efficiency for a Cryogenically Cooled 2.8-GHz, Printed Circuit, 8-Element Antenna Array Using YBCO Thick Films

Temperature (K)	Relative Gain (dB)	Efficiency (%)
300	8.2	20
100	9.2	45
77	9.9	85
50	10.2	88
20	10.5	90

TABLE 7.9 Gain Capability of Two Feed Methods as a Function of Temperature

Temperature (K)	Direct Coupled	Gap Coupled
300	9.0	9.1
77	9.9	10.2
50	10.2	10.6
20	10.4	10.8

7.3 MICROWAVE FILTERS

Superconductor technology significantly improves the performance of microwave filters in terms of insertion loss, skirt selectivity, group delay, and power-handling capability. Whether thin-film or thick-film technology should be used depends strictly on the power-handling capability of the filter.

7.3.1 Low-Power Microwave Filters

Low-power, dielectric filters using superconducting technology (Figure 7.10) are most suitable for communication systems. The significant improvement in the insertion loss is due to improved unloaded Q of the resonators at 77 K. The improvement in the insertion loss of cryogenically cooled 4-pole and 8-pole filters and equalizer filters due to integration of superconducting technology is evident from the tabulated data shown in Fig. 7.10. The improved performance of externally equalized microwave filters demonstrates the feasibility of using superconducting technology in the building of microwave filters for satellites. Integration of both dielectric loaded multiplexing technology and superconducting technology [7] will significantly reduce the insertion loss, weight, and size of the microwave filters used by satellite communication systems.

7.3.2 High-Power Microwave Filters

High-power microwave filters benefit the most from superconducting technology [8]. Such filters are most attractive for microwave transmitter applications. The power-handling capability is contingent on the bandwidth, the unloaded Q of resonators, the filter stages, and the operating temperature. However, improvement in power-handling capability is primarily due to significant improvement in the thermal performance of the filter structure at cryogenic temperatures, the quality of HTSC films, and the superconducting materials.

The thermal quality of the HTSC material plays a key role in defining the upper limit of power-handling capability. Unpatterned YBCO films can handle as much power as unpatterned TBCCO films at 77 K. The insertion loss in cryogenically cooled filters using thick films increases slightly as the input power is increased from 10 to 50 W. The power-handling capability of a 3-pole, HTSC thin-film filter improves as the operating temperature is reduced (Table 7.10).

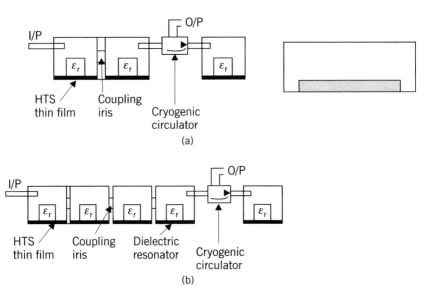

FIGURE 7.10 Layout of (a) 4-pole, superconducting, hybrid, dielectric filter; (b) 8-pole superconducting, equalizer, dielectric filter. (c) Performance data on superconducting dielectric filters at 77 K [7].

Parameter	4-pole HTS Filter	4-pole HTS Filter, Equalized	8-pole HTS Filter	8-pole HTS Filter, Equalized
Parameter bandwidth	1.0%	1.0%	0.8%	0.8%
Minimum insertion loss (dB)	0.12	1.46	0.56	2.21
Loss variation (dB)	0.15	0.2	0.25	0.35
Group-delay variation over 90% of bandwidth (ns)	14	1.5	45	6

(c)

TABLE 7.10 Power-Handling Capability of HTSC Filter Using Thin Films

Temperature (K)	Maximum Input Power (W)
77	2
65	3
20	7
4.2	11

In high-power HTSC microwave filters, passive inter modulation products (PIMPs) are caused by the formation of very thin-film oxide layers on the metal surfaces and mechanical imperfections on the joints, and can present a problem. Microcracks and voids in metal structures produce nonlinearity in the filter, which can cause wide variations in insertion loss and group delays (Figure 7.10).

7.3.3 Advantages of Using HTSC Technology

Implementation of superconducting technology in microwave filters will eliminate the high insertion loss and large size problems generally experienced by high-power microwave filters. Reduction in insertion loss, weight, and size are most desirable for airborne EW receivers, communication satellites, and space sensors. Combining monolithic microwave integrated circuit (MMIC) technology and HTSC superconductor technology will provide narrow-band, high-power filters with minimum loss, flat passband response, and smaller size, which are desirable for airborne sensors. Superconductor technology offers narrowband filtering at microwave frequencies with minimum loss and group delay, thereby eliminating the need for down converters. Thus, HTSC technology can provide improved sensitivity and dynamic range with significant reduction in system cost, weight, and size.

7.4 DELAY LINES

Microwave delay lines are widely used in EW systems for range deception, instantaneous frequency measurement (IFM) receivers, satellite communication transponders, and radar systems. Switching delay lines are used in true time-delay phase shifters for precision electronic beam-steering applications. Conventional delay lines using coax or waveguide transmission lines suffer from high insertion loss, large weight, and enormous size and often require rf amplifiers to compensate for high microwave losses. Conventional delay lines, whether dispersive or nondispersive, limit the dynamic range of the EW receivers in addition to the use of costly rf amplifiers. The cryogenically cooled dispersive delay line commonly known as CHIRP offers wideband frequency compensation for EW compressive receivers and spectrum analyzers. Insertion loss, weight, and size parameters for a 10-GHz coax delay line and a 10-GHz HTSC delay line using superconductor technology are shown in Figure 7.11.

7.4.1 Wideband-Printed Circuit (WPC) Delay Lines

Delay lines using printed-circuit and HTSC technology can offer minimum weight, size, and insertion loss at MM-wave frequencies. Wideband, non-dispersive, delay lines using thin films of HTSC YBCO films on lanthanum

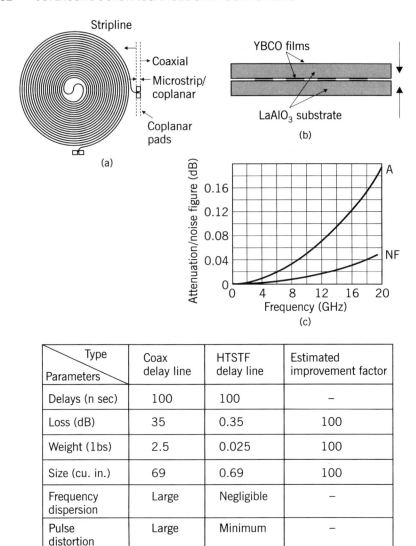

FIGURE 7.11 Physical layout and performance data at 77 K. (a) Nondispersive, superconducting, 22-ns, 20-GHz delay line using YBCO on LaAlO$_3$ substrate and spiral strip line pattern [9]; (b) cross section of assembly; (c) attenuation and noise figure (dB). (d) Comparison between coax and HTSTF delay lines at 10 GHz [9].

aluminate substrate have been designed to operate in MM-wave regions. A 20-GHz, nondispersive, wideband, HTSC delay is shown in Figure 7.11. Variation in insertion loss and noise figure (NF) in a nondispersive, 22-ns, 20-GHz, HTSC-stripline delay line as a function of frequency is illustrated in Figure 7.11 along with a cross-sectional view of the stripline delay line assembly showing the HTSC YBCO films on LaAlO$_3$ [9]. This delay line uses stripline transmission line made of two lanthanum aluminate substrate sections with YBCO thin films. The HTSC YBCO thin film on one side is a pattern in the form of a spiral structure to achieve TEM-field configuration in compact package. This delay line structure offers better overall performance than other planar geometries such as microstrip or coplanar transmission line. The low insertion loss over the entire 20-GHz bandwidth is possible only with HTSC delay lines. Microwave delay lines using superconductor technology not only offer minimum insertion loss, but also consume much less power than conventional coax delay lines when used in a radar system.

7.5 CRYOGENICALLY COOLED FREQUENCY MULTIPLIERS

Low loss varactor diodes are widely used by the solid-state low- and medium-power frequency multipliers operating in lower microwave regions. However, a single-junction varactor diode will not be able to provide moderate power at MM-wave frequencies because of high losses and lower conversion efficiencies at room temperatures. Both the power output and the conversion efficiency improve with multiple-diode junction structure and lower operating temperature.

7.5.1 Conversion Efficiency

The conversion efficiency of a varactor frequency multiplier is strictly dependent on the cutoff frequency or quality factor of the varactor. The cutoff frequency (f_c) is inversely proportional to the series resistance (R_s) of the diode, which is responsible for losses in the junction. Higher conversion efficiency and output power levels are possible with cryogenically cooled frequency multipliers using epitaxially stacked varactor diodes. A cw power output of 9 W at 22 GHz, 5.5 W at 35 GHz, 5 W at 40 GHz, and 250 mW at 94 GHz has been achieved with conversion efficiency of 68, 60, 55 and 12%, respectively, for frequency doublers. Maximum conversion efficiency for various multipliers as a function of operating temperature is summarized in Table 7.11. These calculated values assumed an input frequency of 10 GHz in each case. Higher power levels are possible at lower cryogenic temperatures.

Lower operating temperatures not only improve conversion efficiency and output power, but also improve the heat sink efficiency of a single-junction varactor diode. This will realize considerable reduction in the power dissipation in the junction, which will improve the reliability and the overall conversion efficiency. Higher power levels and conversion efficiencies are possible at cryogenic temperatures using multijunction varactor diodes.

A multijunction varactor diode uses an integrated series IMPATT structure (ISIS), which has a power-handling capability of N^2 times the power capability of a single-junction diode, where N is the number of junction stacks in the ISIS

TABLE 7.11 Conversion Efficiencies for Various Multipliers as a Function of Operating Temperature (%)

Multiplier Type (cutoff frequency, GHz)	Temperature	
	300 K	77 K
Doubler (1000 GHz)	39	87
Trippler (500 GHz)	30	53
Quadrupler (800 GHz)	23	42
Quintupler (1000 GHz)	7	21
Sextupler (1500 GHz)	18	29

structure. The ISIS diodes when operated at cryogenic temperatures can offer the highest power levels at MM-wave frequencies with optimum conversion efficiency. For example, 44-GHz frequency doubler using an ISIS diode demonstrated a conversion efficiency of 52, 68, and 73% at 300, 77, and 40 K, respectively.

The phase noise of the frequency multiplier at room temperature is degraded by an amount equal to $20 \log M$, where M is the multiplying number. This degradation is significantly reduced at cryogenic temperatures under cw operations. Under pulsed conditions, the intrapulse phase change across a 10-μs pulse is less than 2.5° at 300 K, which is reduced to less than a degree at 77 K.

7.6 CRYOGENICALLY COOLED MIXERS

Mixers or down converters are widely used by the radar systems, EW systems, ECM equipment, communication systems, and radiometric sensors. The conversion loss and the intermodulation product (IMP) levels have significant impact on the overall performance of a mixer. The mixer conversion loss and the output frequency affect the dynamic range and noise figure of the receiver, which can limit the detection range of a radar, the sensitivity of an EW receiver, or the fidelity of a communication receiver. Significant improvements in these areas as well as reduction in the local oscillator (LO) power requirement are possible by operating mixers at cryogenic temperatures.

Since high noise figure and conversion loss are generally associated with mixers operating in MM-wave regions, maximum benefit can be expected from a cryogenically cooled mixer operating at MM-wave frequency. By combining HTSC technology and conventional rf components shown in Figure 7.12 in one package and then cooling the entire assembly, one can achieve maximum benefit in terms of low conversion loss, spectral purity, and low LO power requirement. By cooling only the shaded rf components, one can realize substantial improvement in conversion loss, noise figure, LO power requirement, spectral purity, and output noise suppression.

7.6 CRYOGENICALLY COOLED MIXERS   165

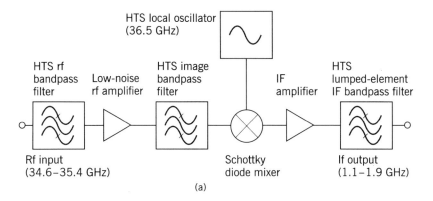

Description of the component	Frequency (GHz)	Temperature (K)
Bandpass image filter	34.6–35.4	77
Bandpass image filter	34.6–35.4	77
Local oscillator	36.5	77
Schottky diode mixer	1.1–1.9	77
Luped-element If filter	1.1–1.9	77

(b)

FIGURE 7.12 Cryogenically cooled 35-GHz down converter using superconducting and conventional components: (a) block diagram; (b) operating frequency and temperature of components.

Performance improvement due to cryogenic temperatures is not limited only to a 35-GHz mixer. The performance data summarized in Table 7.12 on commercially available mixers in various MM-wave frequency bands reveal that cryogenic operations offer significant improvements in noise figure and LO power requirement. The data clearly show a trend in lower double sideband

TABLE 7.12 DBS Noise Figure and Local Oscillator (LO) Power Requirements for Various Cryogenically Cooled MM-Wave Mixers

Frequency Range (GHz)	300 K		20 K	
	NF (dB)	LO (dBm)	NF (dB)	LO (dBm)
75–110	5.08	0.60	2.68	0.20
90–140	5.51	0.60	3.06	0.20
110–170	6.10	0.70	3.41	0.25
140–220	6.45	0.90	4.04	0.30

(DBS) noise figure (NF) and LO power requirement at 20 K operating temperature. Further improvement is possible if the operating temperature is reduced to 4.2 K. The cost of a MM-wave LO source is very high, the phase noise associated with a MM-wave LO is poor, and the power available from a MM-wave LO source is very limited. These limitations and constraints are eliminated or reduced at cryogenic operations.

7.7 CRYOGENICALLY COOLED RF SOURCES

The frequency stability and phase noise of an rf source or microwave oscillator are of paramount importance, regardless of whether it is a solid-state source, a surface acoustic wave (SAW) oscillator, or a crystal oscillator. The parameters are dependent on the operating frequency, resonator structure, quality factor of the resonator, and operating temperature. Significant improvement in frequency stability and phase noise has been observed at liquid helium temperature (4.2 K). HTSC resonators are widely used in frequency synthesizers, where high frequency stability and low phase noise are the principal requirements. Reduction in frequency instability of 15 to 20 dB can have significant effects on radar system performance and mission effectiveness. Coherent radar systems, high-resolution sensors, and advanced ECM systems will benefit the most from cryogenically cooled rf sources.

7.7.1 Rf Oscillators Using Superconducting Ring Resonators

As stated earlier, frequency stability and phase noise are dependent on the resonator. Rf oscillators made from HTSC-sapphire resonators using thin films of YBCO on TlBaCaCuO (TBCCO) substrate offer Q better than 10^6 at 77 K and 10^7 at 4.2 K. However, a ring resonator configuration offers the best overall performance with minimum size and weight. Superconducting microstrip ring resonators [10] operating at X-band frequencies have demonstrated intrinsic Qs better than 7500 at 77 K and better than 20,000 at 25 K. Lower insertion loss is possible only with ultra high Q resonators, which can be realized only at cryogenic temperatures.

Improvement in the frequency stability, insertion loss, and unloaded Q of a X-band superconducting ring resonator oscillator is evident from Figure 7.13. A X-band superconducting ring resonator oscillator using high-quality thin films of YBCO on lanthanum aluminate substrate offers residual frequency stability better than 0.1% at 70 K. Significant improvement in phase noise and frequency stability of MM-wave oscillators are possible only with superconducting ring resonators of Q better than 10^7 at 40 K. These improvements are due to a decrease in London penetration depth as the operating temperature is reduced and the optimum thickness of the superconducting films is used. Theoretical studies indicate that for optimum oscillator performance the super-

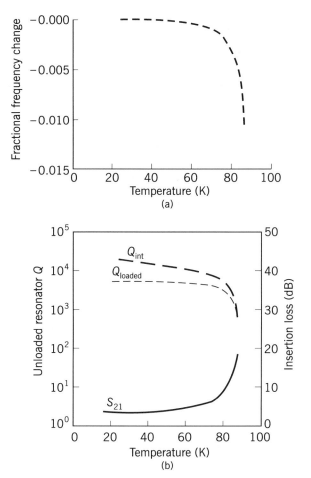

FIGURE 7.13 X-band, superconducting ring resonator oscillator [10]: (a) fractional frequency change with temperature over 4.75–9.5 GHz; (b) impact of temperature on resonator Q and insertion loss.

conducting film thickness must be much less than the effective penetration depth $\lambda(T)$, which is defined by the two-fluid model as follows:

$$\lambda(T) = \frac{\lambda(0)}{1 - (T/T_c)^4} \tag{7.4}$$

where λ_0 = penetration depth at $T = 0$
T_c = critical temperature of the superconducting film

168 SUPERCONDUCTOR TECHNOLOGY IN COMPONENTS

7.7.2 Performance Capability Using Solid-State Devices

This section identifies the improvements in the phase noise and output power level of cryogenically cooled rf sources using solid-state devices such as MESFETs, p-HEMTs, and SL-HEMTs. The low frequency is converted into phase noise at microwave and MM-wave frequencies by the nonlinear parameters of the device. This noise can be reduced by lowering the operating temperature. Both the phase noise and power output of an rf source improve with the decrease in operating temperature, regardless of the type of solid-state device used by the source.

The data in Table 7.13 indicate improvement in both parameters for all three rf sources as the operating temperature is reduced from 300 to 100 K. The improvement in phase noise is strictly due to higher resonator Q at the cryogenic temperatures irrespective of the operating frequency. Further improvement is possible if the operating temperature is reduced to less than 40 K. The above phase noise value is at 10 kHz offset from the 4.6-GHz carrier. Improvement due to superconductor technology is not limited to rf sources. The single-side-band (SSB) phase noise limits for both the solid-state amplifier and the solid-state oscillator improve as the operating temperature is reduced to 77 K or less. Typical phase noise values of cryogenically cooled solid-state amplifiers and oscillators operating in the X-band are summarized in Table 7.14. These values will further improve with the lower cryogenic temperatures. The phase noise will increase by 10 to 20 dB if the operating temperature is increased to 300 K.

TABLE 7.13 Typical Phase Noise and Power Output of Cryogenically Cooled Rf Sources Using Various Solid-State Devices

Parameter	MESFET		p-HEMT		SL-HEMT	
	300 K	100 K	300 K	100 K	300 K	100 K
Phase noise (dB/Hz)	−92	−96	−86	−101	−77	−88
Power output (dBm)	+16	+19	+13.5	+15.2	+10.5	+13.2

TABLE 7.14 Typical SSB Phase Noise for Solid-State Amplifiers and Oscillators at 77 K (dBc/Hz)

Offset (Hz)	Solid-State Amplifier	Solid-State Oscillator
1	−108.5	−102.6
10	−121.3	−115.2
100	−133.5	−127.7
1000	−146.3	−140.6

7.7.3 Capability of Superconducting Rf Oscillator

Hybrid oscillator design using planar superconductive microwave integrated circuit (PSMIC) technology (Figure 7.14) offers remarkable performance in terms of phase noise, power output, and harmonic suppression. This design uses a ring resonator made from YBCO film deposited on $LaAlO_3$ substrate, a GaAs MESFET oscillator, and a transmission line section fabricated from YBCO thin film [11]. A 10-GHz oscillator using PSMIC technology demonstrated output power greater than +11 dBm, efficiency exceeding 12%, second-harmonic suppression better than −35 dBc, and phase noise as low as −68 dBc/Hz at an offset of 10 kHz from the carrier, when cooled down to 77 K. Phase noise of this oscillator was roughly 12 dB lower than that of a copper oscillator at 77 K, while the room-temperature (300-K) phase noise of the copper oscillator was greater than 25 dB compared to that of the superconducting oscillator at 77-K operating temperature under same frequency and offset conditions. Further improvement is possible with this design at lower cryogenic tempera-

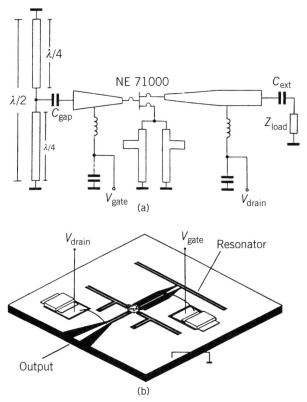

FIGURE 7.14 Physical layout and iosmetric view of 6.5 GHz, superconducting, low-noise oscillator using YBCO films and GaAs devices: (a) circuit diagram of a 6.5-GHz superconducting oscillator using microwave integrated circuit technology [11]; (b) isometric view.

tures in terms of phase noise, frequency stability, output power, efficiency, and harmonic suppression.

7.8 MM-WAVE TRAVELING WAVE TUBE AMPLIFIER (TWTA) DESIGN

Circuit losses and power dissipation in an MM-wave TWTA are extremely high, which will reduce the rf power output and the overall TWTA efficiency. Increase in power dissipation and circuit losses will lead to poor heat transfer efficiency, which will adversely affect the tube reliability. Implementation of the superconductor technology in the TWTA design will bring significant improvements in above areas.

Since most of the losses are associated with the helix structure of the TWTA (Figure 7.15), it is logical to implement superconductor technology in the helix circuit. The helix structure can be made from an HTSC film deposited on beryllium oxide to achieve the highest heat sink efficiency. By lowering the temperature of the helix structure to 77 K or less, one can realize significant improvement in TWTA overall efficiency and rf power output. High-power radar systems and ECM equipment will benefit the most from superconducting TWTAs.

7.9 CRYOGENICALLY COOLED MASER SYSTEMS

Masers offers the lowest noise temperatures and, therefore, can be used in radar systems, where minimum noise temperature or figure is the principal requirement. The maser cavity in which the paramagnetic material is located

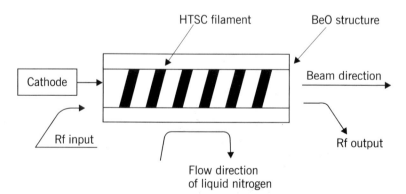

FIGURE 7.15 MM-wave TWT design concept using HTSC technology for improved performance. The helix structure is made from HTSC film deposited on berylium oxide (BeO) substrate because of its high thermal conductivity. The cylindrical structure is also made from BeO. The substrate is externally cooled to 77 K. A depressed collector is used to achieve high efficiency. Berylium oxide acts as a nonelectrically conductive heat sink.

must support both the pump and signal frequencies. The cryostat is placed between the pole pieces of the magnet for maximum sensitivity, as illustrated in Figure 7.16. The cryostat temperature is maintained at 77 K for minimum noise figure of the system. The sensitivity of a maser is dependent on operating temperature and the auxiliary components such as transmission lines and other microwave plumbing elements. The effective noise temperature of a maser can be as low as 25 K, corresponding to a system noise figure of 0.3 dB, which is considered to be the most desirable value for a receiver using a maser rf amplifier. If an additional loss of 0.5 dB is assumed somewhere in the system, the overall system noise figure comes to 0.8 dB, which begins to approach the noise figure of a parametric amplifier. Thus, the use of superconductor technology in a maser will provide the lowest noise figure of the system, which will significantly improve the detection range with minimum cost.

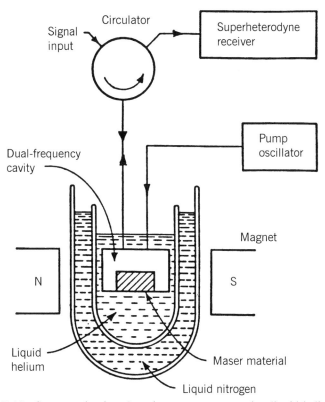

FIGURE 7.16 Superconducting, 3-cavity maser system using liquid helium (4.2 K).

7.10 SUPERCONDUCTIVE FREE ELECTRON MASER SYSTEM

A superconductive free electron maser (SFEM) is capable of providing rf peak power exceeding 200 kW, minimum gain of 30 dB, and rf efficiency as high as 45% in millimeter-wave and submillimeter-wave regions. However, optimum performance of a SFEM system is strictly dependent on an operating temperature much lower than 77 K. Maximum performance improvement is expected if the operating temperature is reduced in the 1- to 4-K range (liquid helium temperature range). An SFEM can be used as an ultralow-noise amplifier. It can be also operated as an extremely stable oscillator, which can be used as a frequency standard. At optical frequencies, the maser is referred to as a laser. A cryogenically cooled free electron maser (Figure 7.17) is capable of generating significant power levels at MM-wave and SUBMM-wave frequencies with maximum efficiency and highest reliability.

7.11 CRYOGENICALLY COOLED MICROWAVE RECEIVERS

A receiver is the most critical component of a radar system, communication equipment, a space-borne sensor, or an ECM system. In the case of radar, the detection, tracking, and acquisition ranges and their associated accuracies are strictly dependent on the performance capability of the receiver. In brief, the radar system performance capability is contingent on the dynamic range and sensitivity of its receiver, which are dependent on the noise figure of the receiver. Furthermore, the radar must detect targets in the presence of severe clutter and jamming environments. A sensitive receiver requires wide linear dynamic range and ultralow-noise figure, which can be achieved by lowering the oper-

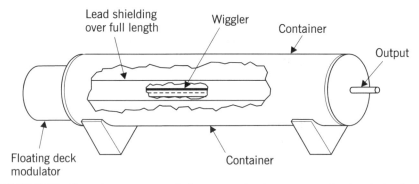

FIGURE 7.17 Free-electron maser design concept using superconducting technology. Projected performance parameters: peak output power capability = 200 kW (maximum), instantaneous bandwidth = 10% (typical), operating frequency range = 10–35 GHz, minimum gain = 30 dB, peak beam current = 10 A (maximum), beam voltage = 400 kV, efficiency = 45% (typical).

ating temperatures of the mixer, local oscillator, amplifier, and other critical elements of the receiver. Improvement in receiver sensitivity due to cryogenic operations can reduce the transmitter power requirement, which in turn can reduce the transmitter cost and power supply weight and size.

7.11.1 Cryogenically Cooled Channelized Receivers

The accuracy of an instantaneous frequency measurement (IFM) system can be significantly improved by using a cryogenically cooled channelized receiver design (Figure 7.18). The analog electronic circuit elements need cryogenic cooling only. The superconducting analog-to-digital (ADC) elements will provide direct synthesis of complex waveforms and digitization of the received signals with high speed and accuracy. The dynamic range of the ADC unit, which is defined as the ratio of the largest clutter signal to system thermal noise, will improve as the operating temperature is reduced. In addition, the losses in the RC-filter elements will decrease as temperature decreases. This means that integration of HTSC technology in a channelized receiver will detect signals even in the presence of clutter and jamming signals over wide

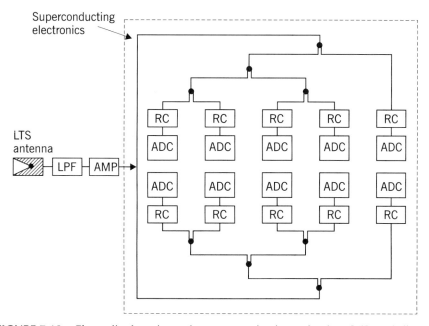

FIGURE 7.18 Channelized receiver using superconducting technology [12]: LTS, linear tapered slot; ADC, analog-to-digital converter; RC, resistance and capacitive elements. Frequency coverage in octave = 2–40 GHz, channel bandwidth = 50 MHz, instantaneous bandwidth = 500 MHz, receiver sensitivity = −60 to −90 dBm, minimum dynamic range = 60 dB, power dissipation = 1.8 W (maximum), channel isolation = excellent.

dynamic range. A typical superconductive channelized receiver offers a channel bandwidth of 50 MHz, instantaneous bandwidth of 500 MHz, receiver sensitivity better than −90 dBm, dynamic range greater than 60 dB, channel isolation more than 50 dB, and power dissipation as low as 1.5 dB, over an octave band in the frequency range of 2 to 40 GHz.

7.11.2 Superconducting Electronic System Measurement (ESM) Receiver

A cryogenically cooled ESM receiver offers significant performance improvement over an uncooled ESM receiver in terms of sensitivity and noise figure, as well as substantial reduction in weight, size, and power consumption. A superconducting ESM receiver offers three times higher instantaneous bandwidth, two orders of magnitude improvement in receiver sensitivity, and about 67% reduction in weight, size, and power consumption. Specific details on performance parameters of cooled and uncooled ESM receivers are summarized in Table 7.15.

7.11.3 Cryogenically Cooled Wideband Compressive Receivers

CHIRP filters form the basis of compressive receivers. Implementation of the CHIRP-transform algorithm in the analog domain is necessary for real-time spectral analysis. HTSC taped delay line CHIRP filters are best suited for instantaneous bandwidth greater than 1 GHz. The HTSC CHIRP filters provide dispersive delays as long as 50 ns with multigigahertz bandwidth and time–bandwidth products exceeding 100. Long dispersive HTSC delay line filters have been fabricated using thin films of YBCO on $LaAlO_3$ substrate with bandwidth capability greater than 3 GHz. Incorporation of HTSC CHIRP filters in a conventional compressive receiver (Figure 7.19), operating at the right cryogenic temperature, will offer instantaneous bandwidth much greater than 3 GHz, which represents significant improvement over the 1-GHz limit of an SAW compressive receiver or the 2-GHz limit of an acousto-optic channelized receiver.

7.11.3.1 Performance Capabilities of HTSC Compressive Receivers The HTSC wideband compressive receiver [12] shown in Figure 7.19 performs

TABLE 7.15 Performance Comparison Between Cooled and Uncooled ESM Receivers

Parameters	Cooled Receiver (77 K)	Uncooled Receiver (300 K)
Frequency range (GHz)	2–16	2–20
Instantaneous bandwidth (GHz)	1.5	0.5
Maximum sensitivity (dBm)	−90	−60
Relative weight and size (%) reduction	33 (ea)	100 (ea)

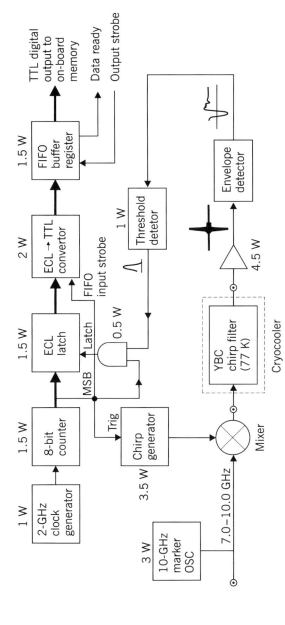

FIGURE 7.19 Architecture of a compressive receiver using cryogenically cooled chirp filters at 77 K [12].

spectral analysis in real time with 100% time coverage. It is suitable for military electronic warfare and molecular spectroscopic applications. No other receiver architecture comes close to an HTSC compressive receiver configuration. Its frequency resolution and processing capability (Table 7.16) are superior to those of an SAW compressive receiver. Advantages of this superconductive receiver include low loss microwave circuits and nondispersive penetration depth, which lead to a frequency dispersive-free performance.

The HTSC compressive receiver offers improved sensitivity, amplitude accuracy, frequency resolution, time-of-arrival (TOA), short-pulse capability, and dynamic range compared to an SAW compressive receiver. Furthermore, the processing power, indicated by the equivalent digital fast Fourier transform (FFT) process in floating-point operation per second (FLOPS), is most impressive. Minimum power consumption is another significant advantage of

TABLE 7.16A Performance Capabilities of Superconducting CHIRP Filters [12]

	Frequency Accuracy (MHz)	Minimum Number of Filters	Equivalent Number of Channelizing Filters
HTS compressive receiver (3-GHz BW)			
Without interpolation	13.3	1	225
With conservative interpolation	3.3	1	900
HTS compressive receiver (10-GHz BW)			
Without interpolation	13.3	1	750
With conservative interpolation	3.3	1	3000

TABLE 7.16B Performance Levels of HTSC and SAW Compressive Receivers (Digital) [12]

	BW (GHz)	N	Duration NT (ns)	Frequency Resolution (MHz)	Equivalent DSP Rate (GFLOPS)
HTS compressive	10.0	2048	102.4	9.8	1320
	3.0	512	85.4	12	324
	3.0	256	42.7	23	288
	3.0	128	21.3	47	252
SAW compressive	1.0	256	128	7.8	96
	0.30	8192	13,660	0.073	47
	0.30	128	213	4.7	25
	0.050	4096	40,960	0.024	7.2

Note. DSP rate to perform an N–point complex FFT:
$$\text{DSP rate (FLOPS)} = \frac{19.92}{NT} N \log N$$
$$= 2 f_{max}(19.92) \log N$$
where f_{max} is the maximum clock frequency.

the HTSC compressive receiver. A 50-MHz bandwidth, 40-µs long analog CHIRP transform algorithm in an SAW compressive receiver with a 300 GFLOPs DSP rate would require a power consumption greater than 4 kW, which is about 10 times the estimate for the HTSC-YBCO compressive receiver. Effective implementation of superconductive technology in compressive receivers requires development of high-quality, low-loss, HTSC thin films, high-speed semiconductor technology, low-cost cryocooler design, and exotic microwave components. Cryocooler cost, reliability, weight, size, and efficiency will continue to improve with the rapid improvement of infrared and HTSC technologies.

7.12 CRYOGENICALLY COOLED MM-WAVE SURVEILLANCE RECEIVERS

Superconducting MM-wave surveillance receivers can play a significant role in military applications, because of their wideband and ultralow-noise capabilities. A cryogenically cooled surveillance receiver offers noise temperature as low as 50 K over the 40- to 60-GHz range and 100 K over the 94- to 110-GHz range. Low single-sideband noise temperature, instantaneous bandwidth in excess of 1.5 GHz, estimated size of less than one-fifth of the conventional uncooled receiver, and estimated weight less than one-tenth of the uncooled receiver are the major advantages of a superconducting surveillance receiver for MM-wave applications.

7.13 SUPERCONDUCTING ANALOG-TO-DIGITAL CONVERTER (ADC)

Cryoelectronics offers significant improvements in the performance of radar receivers, signal processors, and ECM systems. High clutter rejection requires low transmitter phase noise and wide dynamic range of the receiver. Both requirements depend on the ADC capabilities and integration of superconducting technology in relevant components. Integration of LTSC technology in an ADC unit will improve the operational speed of the ADC, the dynamic range of the receiver, and spurious levels generated by the quantization process. A 12-bit, superconducting ADC has demonstrated a power consumption less than 100 µW. The ADC employs several hundreds of Josephson junctions (JJs) made of niobium and aluminium oxide capable of operating at 4.2 K or lower. The quantization of the ADC is a two-junction superconductive quantum interference device (SQUID), which emits a pulse every time the incoming rf signal increments by a quantum level.

An HTSC material must have a bandgap three to five times larger than that of an LTSC material. This means that power consumption in JJ devices made from HTSC materials will be three to five times higher that of devices made from LTSC materials. Therefore, HTSC materials such as YBCO and TaBaCaCuO are not suitable for JJ devices and rapid single flux quantum

(RSFQ) logic circuits [13]. The flux quantum is defined as the ratio of Planck's constant to electron charge and is equal to 2.07×10^{-15} weber. The high speed of JJ devices allows sampling rates in excess of 100 GHz and the RSFQ logic circuits achieve signal processing in real time. This is all possible with the application of LTSC technology to ADC components.

7.13.1 Critical Elements of an ADC

SQUIDs are widely used in the electronic circuits for superconducting ADC devices. A SQUID consists of a superconducting loop or loops that are interrupted by one or more JJs. When a magnetic flux passes through a loop, it produces a phase difference. JJs using only LTSC technology offer fast switching with low dissipation and high sensitivity to electromagnetic (EM) signals. An ADC generally uses one comparator circuit per bit, employing a one-junction SQUID, the dynamics of which are less susceptible to distortion. The ADC comparator circuit using a quasi-one-junction SQUID shown in Figure 7.20 is most attractive, because it is capable of being clocked at 20 GHz with 4-bit conversion of a 10-GHz sinewave signal. Potential SQUIDs and comparators circuits for possible application to superconducting ADCs are shown in Figure 7.20.

Logic circuits employing inductive coupling are best suited to suppress the critical current. This condition is necessary for optimum performance of a superconducting ADC. Since the JJ device is symmetrical, it can provide logic functions with gate currents of either polarity. Potential logic circuits that can be used in superconducting ADCs are shown in Figure 7.21.

Logic circuits made from HTSC materials and operated at higher cryogenic temperatures will have to use higher currents, leading to higher noise levels. That is why logic circuits for superconducting ADCs must have LTSC-JJ device technology for optimum performance.

The sampler circuit is the heart of a sampling system. High sensitivity and high speed are necessary for high-performance ADCs, which are possible only with HTSC technology. A typical circuit configuration of an HTSC sampling device consists of three circuit elements: a pulser, a comparator, and an output gate. This sampler design is a nonlatching comparator junction whose switching generates a circulating current in a superconducting loop. Trapping of this current must be avoided to achieve reliable operation of the ADC.

7.13.2 Performance of a High-Resolution, High-Sensitivity ADC Using LTSC Technology

As stated earlier, superconductive microelectronic technology offers an ADC design with high sensitivity, reliability, linearity, and wide bandwidth with minimum power consumption. The architecture of a superconductive high-resolution ADC is illustrated in Figure 7.22. Superconducting digital circuit functions can be performed on the ADC chip at rates greater than 20 GHz with

7.13 SUPERCONDUCTING ANALOG-TO-DIGITAL CONVERTER (ADC)

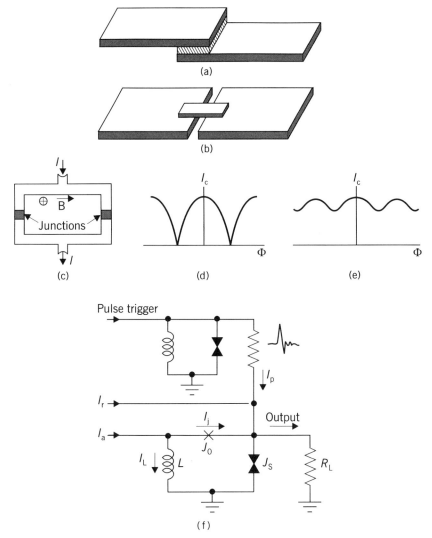

FIGURE 7.20 Josephson-junction architectures for ADC circuits [13]: (a) sandwich-type JJ; (b) coplanar JJ; (c) two-junction SQUID; (d) loop current, I_c, when inductance $L = o/I_{c1}$, and o = flux quantum; (e) loop current, I_c, when inductance $L = o/I_{c1}$; (f) ADC-comparator circuit with one-junction SQUID.

a power density of few mw/cm^2, which represents significant improvement over uncooled ADC devices using GaAs/AlGaAs-HBT processing technology. This particular ADC architecture uses a technique based on the RSFQ logic concept, which utilizes LTSC-JJs. Simple RSFQ circuits developed by HYPRESS, Inc. (NY) have demonstrated clock frequencies exceeding 100 GHz. The innovative architecture developed by HYPRESS for a 10-bit

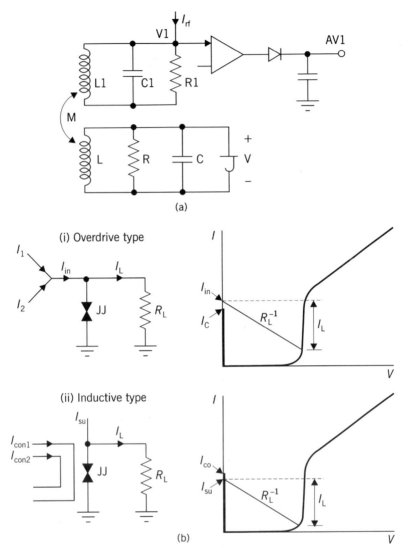

FIGURE 7.21 Circuit diagrams for (a) RF-SQUID and (b) Josephson logic circuits (JJ = Josephson Junction)[13].

ADC unit uses more than 800 JJs and has demonstrated a dynamic range equivalent to that of 14-bit unit. This unique LTSC technology offers high resolution, high data rates, high sensitivity, ultralow-noise, and power consumption less than 5 mw/chip at an operating temperature less than 4.2 K. The ADC architecture can average and decimate the data to an optimum sampling rate, which will increase dynamic range and reduce harmonic-related spurious levels.

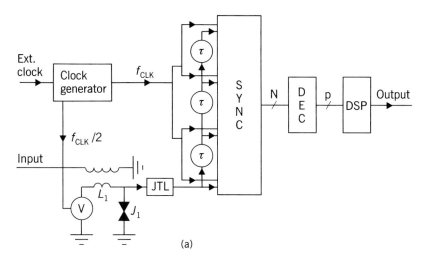

FIGURE 7.22 (a) High-resolution ADC using a rapid single-flux quantum (RSFQ) device; (b) performance levels of HTSC-ADC devices [13].

7.13.3 Superconducting Sigma–Delta ADC Architecture Using Single Flux Quantum (SFQ) Technique

Sigma-delta ADCs have dominated the market for high-performance systems that demand a linearity of 16-bits or more. This performance level is possible through a combination of oversampling and feedback. Superconductive SFQ technology is an excellent candidate for extending the bandwidth of ADC devices, which is required by advanced high-performance radars and ECM systems. Precision feedback is the key to the performance of LTSC sigma/delta ADC devices. Niobium-based LTSC technology when implemented in the design of a 10-bit ADC offers spurious-free dynamic range in excess of 78 dB and third-order intermodulation products (TOIMPs) better than −68 dB, based on computer analysis. Computer analysis further indicates that the extension of flux quantization feedback theory to the multiple loop concept will further improve the dynamic range and instantaneous

182 SUPERCONDUCTOR TECHNOLOGY IN COMPONENTS

bandwidth of the ADC devices using LTSC technology. Flux quantization provides a quantum mechanically precise feedback mechanism that is not available by other means. Superconducting high-resolution ADC design based on LTSC technology appears to be the only way to achieve simultaneously low aperture time of the front end, high internal clock frequency, high reliability, high data rates over wideband, and high static accuracy with minimum power consumption.

7.13.4 Superconducting Digital Radio-Frequency Memory (DRFM)

A DRFM-based system can simulate the performance of advanced radars using complex waveforms and sophisticated EW systems employing advanced ECM techniques. A block diagram of a wideband DRFM system is depicted in Figure 7.23, which uses superconducting components enclosed in the dotted rectangle. Only the critical components need to be maintained at cryogenic temperatures not exceeding 77 K. The superconducting DRFM system allows the ECM techniques to employ deception modulation concepts on the replica of the radar waveform at the optimum time for maximum ECM effectiveness.

Low spurious levels over wide instantaneous bandwidth is the principal requirement for a DRFM-based system, a condition that cannot be met by the uncooled components used by the system. The ADC device is the heart of a

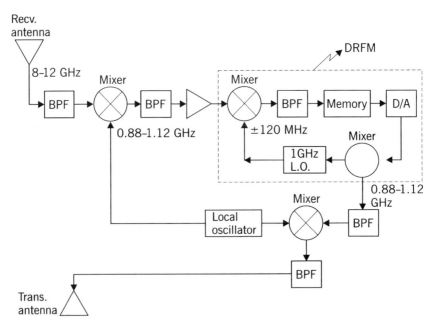

FIGURE 7.23 Architecture of an rf digital memory (DRFM) device using superconducting components (broken line rectangle).

DRFM system and an uncooled ADC will not be able to meet the stringent system requirements. However, incorporating niobium-based RSFQ superconductive logic in the ADC architecture will allow practically all the DRFM system performance requirements to be met. The high-accuracy requirement for high resolution can be satisfied only with cryogenically cooled fast electronic circuits having time constants on the order of the 2 to 4 ps. Fast switching speeds and setting the right threshold levels are possible only through niobium-based superconducting ADC devices.

A complete superconducting 20-GHz DRFM system operating at over 10-GHz instantaneous bandwidth is feasible with ADC, DAC, and 64 memory chips integrated on a single CHIP. The entire system can fit on a 6-inch wafer with power dissipation less than 500 mW under cryogenic operations. Such a DRFM system has potential application in a strategic platform.

7.14 SUPERCONDUCTING PACKET SWITCHES FOR COMMUNICATION SYSTEMS

Very broadband throughputs in excess of 1000 GBPS are required for the asynchronous transfer mode (ATM) in heavily loaded communication systems. A superconducting packet switch is capable of meeting these requirements. The switch design requires the use of superconducting technology in various architectures, including shared memory types, shared bus types, space-division Banyan types, and space-division crossbar types. Trade-off studies performed on various switch architectures indicate that a superconducting packet switch of space-division Banyan-type architecture gives optimum performance at the right cryogenic temperature. Broadband superconducting stripline structure and a superconducting network system can achieve high throughput rates that are not possible with uncooled packet switching systems.

A superconducting switch design consists of an input buffer, a contention element, and a distribution network. The distribution network distributes contention-free packets to the final destination in minimum time. Current HTSC technology indicates that 4-GHz clock operations are feasible, which will provide a maximum throughput of 64 GBPS. Generally, a minimum of five clock cycles are required for a packet to pass through a switch, which indicates a minimum turnaround time of 1.25 ns. A superconducting packet switch fabricated with a standard niobium trilayer process provides a JJ with a critical current density of 2800 A/cm^2 at 4.2 K. A SQUID driven by a three-phase clock can also be used in the switch. Potential applications of this switch include public gigabit networks and ATM data exchanges in public communication systems.

7.15 CELLULAR-BASE STATIONS

High-power microwave filters using HTSC thin films of YBCO on lanthanum aluminate substrate can offer significantly improved performance of cellular-base stations. The HTSC forward-coupled microstrip filters have demonstrated low insertion loss and high power-handling capability over the communication bandwidth with minimum interference. The input power is limited by thermal dissipation in the microwave filter structure, which is significantly reduced at cryogenic temperatures not exceeding 77 K. A 5-pole, 2-GHz filter demonstrated an insertion loss of less than 0.25 dB and power-handling capability of 27 W (cw) at an operating temperature of 10 K, which is impossible at room temperature. The filter performance can be further improved by selecting a superconducting material with lower critical temperature and operating at a much lower cryogenic temperature.

7.16 SPACE COMMUNICATION SYSTEMS

HTSC technology can play a key role in improving the information data rate of a space communication system. The data rate of communication is dependent on an effective radiated power (ERP) level (the product of transmitter power output and transmitter antenna gain), range and operating temperature. Data rate can be expressed as

$$D = \left[\frac{P_t G_t A}{R^2}\right] \bigg/ \left[(4\pi)\left(\frac{C}{N}\right)(kT_o\beta)\right] \tag{7.10}$$

where D = data rate
β = communication parameter (assumed 10)
T_o = operating temperature (K)
R = range (m)
C/N = carrier-to-noise ratio
A = receiver antenna area (square meters)
k = Boltzmann's constant (Joules/K)
P_t, G_t = transmitter power (W) and transmitter antenna gain.

It is evident from Equation 7.10 that the data rate can be increased by lowering the operating temperature or increasing the transmitter power, which is a costly option, or by increasing the receiver antenna aperture, again a costly option. The most cost-effective approach for increasing the data rate is obviously to operate at a cryogenic temperature of 77 K or less. Improved data rates as a function of ERP, antenna aperture, and range for various space communication systems have been observed at 77 K. Data rates can be further increased at least by a factor of 18:1 from room temperature if the operating temperature is to reduce to 4.2 K. Cryogenic operation will not only increase the data rates of space-borne communication systems, but will also reduce the transmitter power requirement, thereby realizing substantial reduction in system weight, size, power consumption, and launching cost.

7.17 RADAR SYSTEMS

A radar system uses several costly components and subsystems, such as a high power transmitter, transmit and receive antennas, mixers, filters, and digital and analog signal processors. The performance of most of these elements can be significantly improved through the integration of HTSC and LTSC superconducting technologies. The application of superconducting technology to most of these components has been discussed previously. The implementation of HTSC and LTSC technologies to certain specific radar components will improve detection and tracking capabilities under severe clutter and ECM environments.

7.17.1 Cryogenically Cooled FM–CW Radar Capability

A block diagram of a cryogenically cooled FM-CW radar system is shown in Figure 7.24 and the components to be cooled are clearly identified. Cooling the delay line, YIG-tuned oscillator, and mixer can significantly improve the radar in terms of clutter rejection, frequency stability, and detection range. Cooling

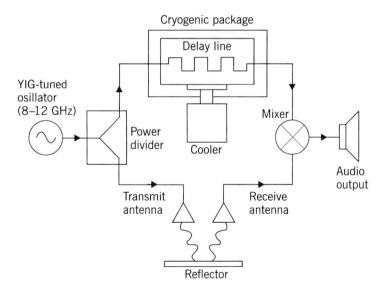

FIGURE 7.24 Block diagram of a cryogenically cooled FM-CW radar device with wideband sweep capability. The cryogenic cooler is mounted on the top of the radar assembly and is designed to keep the superconducting delay line temperature below 77 K.

of each and every component down to 77 K is not necessary. Only the critical components identified above need to be maintained at 77 K. In this particular case, cooling only the delay line section to 77 K gives maximum benefit with minimum cost and complexity. Cooling the delay line down to 77 K reduces the weight, size, and power consumption by an approximate factor of 5:1.

7.17.2 Phased-Array Radar Systems

Critical components of a cryogenically cooled phased-array radar system are shown in Figure 7.25. The cost of an electronically steerable, phased-array radar system is much higher than that of a mechanically scanned radar, because of a large number of costly phase-shifting elements with drivers and power amplifiers are involved. So even by cooling the selected components such as phase shifters, power amplifiers, driver control circuits, and T/R switches, one can reduce the overall insertion loss and power consumption in the radar system. Components with high room-temperature losses must be considered for cooling, because of the possibility of substantial reduction in the transmitter output requirement. The room-temperature insertion loss in an X-band, 4-bit, PIN-diode phase shifter is about 2.5 dB, which can be reduced to less than 1 dB by cooling it to 77 K. The room-temperature loss in an X-band T/R switch is about 2 dB, which can be reduced to less than 0.8 dB at 77 K. Similarly, the losses in the passive X-band components can be reduced at least by a factor of 2:1 at 77 K. One can expect significant reduction of insertion loss in the complex antenna feed assembly at 77 K, because of the large number of waveguide, coax, and stripline components involved.

Data handling capability and processing power of radar signal processors will improve at cryogenic temperatures. Studies must be undertaken to determine which components need to be cryogenically cooled to satisfy cost-effective criteria. Based on experience, maximum benefit from the integration of HTSC technology in a phased-array radar is possible by cooling only the critical components, such as phase shifters, delay lines, ADC devices, and a few others, to minimize the overall system cost and complexity.

ADC devices must use LTSC technology to derive maximum benefit, while the phase shifters, delay lines, mixers, and filters can use the less expensive HTSC technology. For cryocoolers with limited heat capacities, the thermal mass of the components must be kept low. Heat flow into the system from the surroundings must be carefully investigated for the radar components to be operated at cryogenic temperatures. Finally, there is compelling evidence that integration of superconductor technology in selected radar components will not only bring significant improvements in the radar performance and reliability, but also will reduce the transmitter output and input power requirements, leading to a remarkable reduction in system procurement cost.

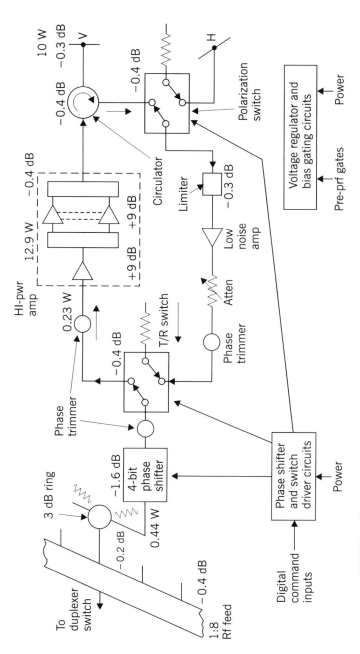

FIGURE 7.25 Application of superconducting technology to high-power phased-array components (shaded areas). Shaded components require cooling at 77 K. H = horizontal polarization, V = vertical polarization.

SUMMARY

Integration of superconducting technology in microwave components and devices will improve system performance and reduce power consumption, weight, and size. Components such as mixers, antennas, filters, delay lines, frequency multipliers, and A/D converters have demonstrated significant improvement in performance at cryogenic temperatures. When employed by radar systems, space sensors, ECM equipment, and communication systems, cryogenically cooled components will improve reliability, efficiency, stability, and noise level. High-power phased-array radars, airborne ECM equipment and satellite-based high-resolution systems will benefit the most from the cryogenically cooled rf devices and components.

REFERENCES

1. Y. Nelson et al. Cryogenic characteristics of wideband p-HEMT, MMIC low-noise amplifiers. *IEEE Trans. MIT* 41:992–999 (1993).
2. P. Penfield et al. *Varactor Applications*. MIT Press, 1962, p. 178.
3. N. Suzuki et al. Properties of an electromagnetically coupled, small antenna with a superconducting thin-film radiator. *IEEE Trans. Appl. Superconductivity* 6:13–17 (1996).
4. L. P. Ivrissimtzis et al. On the design and performance of electrically small, printed-circuit, thick-film YBCO antennas. *IEEE Trans. Appl. Superconductivity* 4:33 (1994).
5. L. P. Ivrisstmzis et al. High-gain, series-fed printed-circuit dipole arrays of high T_c superconductors. *IEEE Trans. Antennas Propag.* 42:1419–1426 (1994).
6. M. A. Richard et al. Microstrip antenna: an experimental comparison of two feeding methods. *IEEE Trans Antennas Propag.* 41:967–973 (1993).
7. R. R. Mansour et al. A C-band superconductive input multiplexer for communication satellites. *IEEE Trans. MTT* 42:2472–2478 (1994).
8. R. R. Mansour et al. On the power-handling capacity of the high-temperature superconductive filter. *IEEE Trans. MTT* 44:1322–1333 (1996).
9. S. H. Talisa et al. High-temperature superconducting wideband delay lines. *Microwave J.*, 88–96 (1995).
10. Electrical characteristics of thin-film $BaYCuO_7$ superconducting ring resonators. *IEEE Microwave and Guided Wave Letters* 1(3) (1991).
11. N. J. Rohrer et al. Hybrid high-temperature superconductor/GaAs 10-GHz microwave oscillator: temperature and bias effects. *IEEE Trans. MTT* 41:1865–1871 (1993).
12. W. G. Lyon et al. High-temperature superconductive wideband compressive receivers. *IEEE Trans. MTT* 44:1258–1275 (1996).
13. T. V. Duzer. Superconductor electronic device applications. *IEEE J. Quantum Electron.* 25: 2365–2376 (1989).

CHAPTER EIGHT

Applications of Superconducting Technology to Electrooptical Components and Systems

This chapter focuses on the application of superconducting technology to electrooptical (EO) components and systems for both military and commercial use. Performance improvement in EO components due to integration of high-temperature and low-temperature technologies is discussed. EO devices and components include optical detectors, IR detectors, photoconductive detector arrays, hot bolometers, solid-state laser diodes, including quantum-well (QW) laser diodes, SIS mixers, IR camera using cryogenically cooled focal plane arrays technology. The EO systems include SIS-quasi-particle receivers, forward-looking IR(FLIR) systems, optical modulators, IR communication transmitters, and superconducting scanning microscopes. The emphasis is on performance improvement and critical issues due to integration of superconducting technology.

8.1 DETECTORS: PHOTON, QUANTUM, AND OPTICAL

This section investigates the performance improvement of photon detectors, quantum detectors, and other optical detectors when operated at cryogenic temperatures. Most photon detectors show significant improvement in performance even at higher cryogenic temperatures. Typical operating temperatures of commercially available photon detectors [1] are listed in Table 8.1. Improvement in detectivity and responsivity as a function of cryogenic operating temperature is summarized for specific detector categories. The Johnson

TABLE 8.1 Operating Temperatures for Commercial Detectors [1]

Detector Material	Temperature (K)
Si	300
Si	300
PbS	300, 193, 77
PbSe	300, 145 to 250, 77
InAs	300, 195, 77
InSb	300, 77
Ge:Au	77
Ge:Hg	5
Ge:Cd	5
Ge:Cu	5
Ge:Zn	5
Si:As and Si:Ga	20
$Hg_{1-x}Cd_xTe$	77
$Pb_{1-x}Sn_xTe$	77

noise, or thermal noise, is generated in the detector due to the temperature rise in its junction. As the temperature of the junction resistance increases, the mean kinetic energy of the carrier increases, leading to increased electrical noise voltage. This type of noise can be significantly reduced at cryogenic temperatures.

Quantum detectors, which are sensitive in the visible and near infrared (IR) spectral regions, will benefit the most at cryogenic operations. Cryogenically cooled quantum detectors will be most suitable for fiber optic-based systems, namely, optical delay lines, optical data links and fiber optic (FO) ring lasers. Besides detectivity and responsivity capabilities, the impact of cryogenic temperature on spectral range, sped, rise and fall times, dark current, noise equivalent power (NEP), and dynamic range are investigated. Some detectors require cooling below 77 K even for normal operation. Quantum-well infrared (QWIR) detectors have to be cooled well below 77 K to provide background-limited performance. Ternary compounds such as mercury–cadmium–telluride (Hg:Cd:Te) require cryogenic cooling to provide acceptable performance for longer wavelength operations.

8.1.1 Performance of Cryogenically Cooled Photon Detectors

Photon detectors include PbS (77 K), InAs (77 K), Ge:Au (77 K), Ge:Hg (5 K), Ge:Cd (5 K), Ge:Cu (5 K), Ge:Zn (5 K), Si:As (20 K), and Hg:Cd:Te (77 K) and require cryogenic cooling at their respective temperatures (specified in parentheses). Theory indicates that a photon detector should have more photon-generated carriers than thermally generated carriers. However, a high-temperature (high-T_c) photon detector optimized to detect room-temperature blackbody radiation will require a much lower operating temperature than a conventional photovoltaic detector. Superconductor technology can play a key role in the design and operation of a photon detector using NbN or YBCO film. Cryogenically cooled detectors will be most suitable in measuring the response of visible and IR laser radiation. When operated at or below 22 K, the detector offers fast response over the 1- to 10-μm spectral range with a detectivity (D^*) better than 10^9 cm$\sqrt{Hz}$/W.

Performance capabilities of some cryogenically cooled photon detectors are shown [1] in Figure 8.1 through 8.4. Figure 8.1 indicates superior performance of a Pb:Se detector even at 193 K, which can be further improved if the operating temperature is reduced below 150 K. Figure 8.2 shows a sharp drop in the detectivity of a Ge:Zn photon detector, if the operating temperature exceeds 10 K. Figures 8.3 and 8.4 illustrate improvement in both detectivity and spectral response over a wide temperature range.

The most sensitive superconducting detectors operating over very wide spectral ranges are summarized in Table 8.2. The extrinsic doped silicon, germanium, and germanium–mercury detectors require operating temperatures below 30 K, employing closed-cycle cryocoolers. These detectors present serious logistic problems for applications where reliable operation over extended periods is the principal requirement. On the other hand, Hg:Cd:Te and Pb:Sn detectors offer good performance and maximum economy even at 77 K operating temperature. The cryogenically cooled Pb:Se and In:S detectors offer high detection capabilities over 3–5 spectral range with cryogenic cooling ranging from 145 to 193 K, thereby providing a flexible cooling scheme and maximum economy. The Pb:Se detector is the most suitable for low-frequency operation because of its low time constant, while the In:S photon detector is most attractive for higher-frequency operation. The performance of both detectors is significantly improved when they are cooled down to 77 K.

8.1.2 Cryogenically Cooled IR Detectors

Ternary alloy detectors, namely, mercury–cadmium–telluride (Hg:Cd:Te) and lead–tin–telluride (Pb:Sn:Te), are widely used as IR detectors because of superior performance even at higher cryogenic temperature (77 K). By using a right partial mole fraction in Hg_{1-x}:Cd_x:Te and by operating at a specific cryogenic temperature, one can achieve an optimum performance over a wide spectral range. The maximum possible detection level of an IR detector is dependent on the energy gap of the material and the operating temperature.

Research activities [2] reveal that a superconducting photo-field-effect transistor (SUFET) acts like a photo detector with improved performance. This three-terminal device SUFET, shown in Figure 8.5, uses a YbACuO/PrBaCuO channel on a transparent superconducting substrate MgO and has potential application in the visible and infrared regions. This detector can operate at 77 K over a 10-kHz frequency range with minimum signal distortion. Higher-frequency operation is possible with channel dimensions down to micrometer range. This particular detector also shows a good response up to optical mod-

192 SUPERCONDUCTING TECHNOLOGY IN ELECTROOPTICAL COMPONENTS

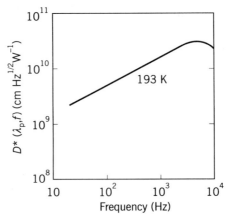

FIGURE 8.1 Capability of the cyrogenically cooled Pb:Se detector [1]: $T_d = 145$–250 K, $A_d = 25 \times 10^{-6}$ to 1 cm^2, $R_d = 2$–100 MΩ, $\tau = 10$–100 μs, FOV $= 2\pi$ sr, background temperature $= 300$ K, $\mathcal{R}_{\lambda p} \approx 10^5$ V/W, $\mathcal{R}_{\lambda p}/\mathcal{R}_{bb} \approx 5$.

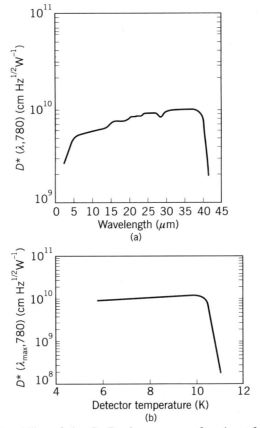

FIGURE 8.2 Capability of the Ge:Zn detector as a function of (a) wavelength and (b) temperature [1].

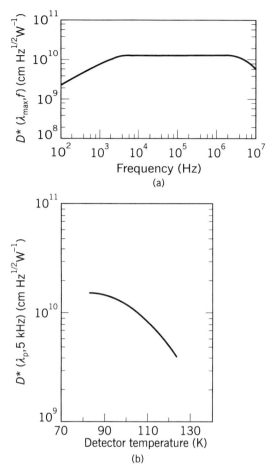

FIGURE 8.3 Capability of the Hg:Cd:Te detector as a function of (a) frequency and (b) temperature [1].

ulation frequency exceeding 1 MHz. Variations in current density as a function of cryogenic temperature can be seen in Figure 8.5.

8.1.3 Quasi-Optical SIS Detectors

A superconductor–insulator–superconductor (SIS) is the best tool for conducting Fourier transform spectrometer studies in the 300- to 1000-GHz frequency range [3]. The behavior of niobium-based SIS junctions and superconducting tuning circuits near the gap frequency is of paramount importance. Niobium (Nb) is a normal metal that operates far above the gap frequency. The electromagnetic field penetration around the gap frequency is clearly defined by the skin depth. However, in the region around the gap frequency, there is a transi-

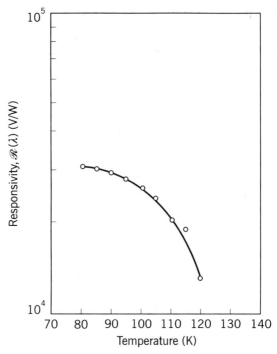

FIGURE 8.4 Sensitivity of the He:Cd:Te detector as a function of temperature [1].

TABLE 8.2 Superconducting Detectors Operating Over Wide Spectral Ranges

Photon Detector	Spectral Range (μm)	Cryogenic Temperature (K)
Ge:Hg	8–14	27
Ge:Cu	8–28	5
Si:As	8–30	5
Si:Ga	8–16	27
He:Cd:Te	8–14	77
Pb:Sn:Te	8–14	77

tion between the superconducting and normal behavior. SIS detector performance is dependent on the gap frequency, skin depth, London penetration depth (λ_0), and the operating temperature. The London penetration depth is given as follows:

$$\lambda_0 = \delta \pi \omega_g \tag{8.1}$$

where ω_g is the angular gap frequency.

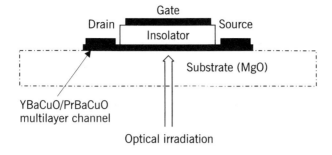

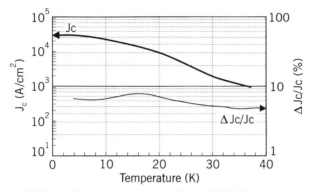

FIGURE 8.5 Schematic view of the SUPFET detector [2].

Maximum penetration depth is obtained at a frequency slightly higher than the gap frequency (Figure 8.6). Modeling of the SIS detector can be achieved by linear approximation of the current–voltage characteristic of the SIS junctions, which are represented by the admittances that are connected through short transmission lines with the impedances. Both SIS junctions are dc-biased in parallel and the junction dimensions are about one-eighth of the incident wavelength at the highest frequency. The performance of an SIS detector is strictly a function of junction capacitance and operating temperature. The optimum performance of Nb-based detectors over the 500- to 700-GHz range has been observed over a very narrow temperature range of 3.1 to 4.2 K [3] and the peak detector response was noticed above the superconducting gap frequency. A quasi-optical, twin-junction SIS detector is capable of detecting electromagnetic signals up to 1000 GHz. Detector response increases as the operating temperature is reduced to 3.1 from 4.2 K. Detector response is greater at higher frequencies. The increase in detector response is about 15% for frequencies below the gap frequency and roughly 50% above the gap frequency.

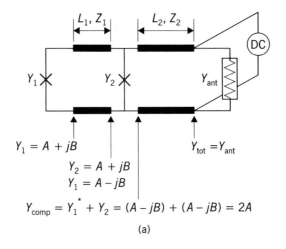

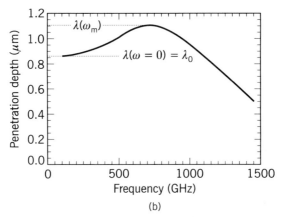

FIGURE 8.6 (a) Parameters for the Nb superconducting SIS junctions [3]. (b) Frequency dependence of penetration depth into Nb junctions.

8.2 SUPERCONDUCTING BOLOMETERS

Basic knowledge of conventional bolometers is essential to understand the performance capabilities of superconducting bolometers. The bolometer effect is a change in the electrical resistance of the responding element due to a change in temperature produced by the absorption of the incident IR radiation. The two electronic circuit configurations shown in Figure 8.7 illustrate the use of this effect. The incident IR radiation will cause a rise in temperature of the detector resistance, which will result in a drop in its resistance, thereby electronically unbalancing the bridge. The change in bolometer resistance is dependent on the temperature coefficient and detector temperature. The temperature coefficient is defined as follows:

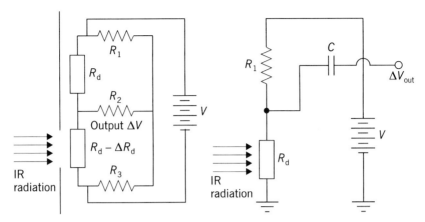

FIGURE 8.7 Potential circuit configurations for a cryogenically cooled bolometer detector: R_1, R_2, and R_3 are resistive elements of the electrical bridge circuit; R_d is the photodetector resistance.

$$\alpha_t = \frac{1}{R_d} \frac{dR_d}{dT_d} \tag{8.2}$$

where R_d = detector resistance
T_d = detector temperature

But the change in detector resistance (ΔR_d) is given as follows:

$$\Delta R_d = \frac{dR_d}{dT_d} \Delta T_d \tag{8.3}$$

where ΔT_d is the temperature change in the detector. Studies performed on the bolometers indicate that both the sensitivity of the temperature coefficient of the resistance and the resposivity improve as the operating temperature of the bolometer is reduced.

Bolometer detectors with a wide range of sensitivities are used from visible wavelengths to microwave frequencies for detection of signals and absolute power measurements. Superconducting bolometers use the temperature dependence of the resistance of a superconducting film near the critical temperature T_c as a thermistor element. Noise equivalent power (NEP) indicates the performance level of a bolometer. The NEP of a bolometer consists of a photon noise component present in the incident radiation, a photon noise component due to the exchange of photons between the bolometer and heat sink, and the Johnson noise in the thermistor. The first two NEP components provide a fundamental limit to wideband bolometer sensitivity with given values of critical temperature T_c and the ambient temperature T_a. The $1/f$ noise component does not present any problem in the superconducting film of high quality. For a sensitive and accurate bolometer readout, the detector responsivity should large enough that the Johnson noise components are not the dominating ones. Superconducting bolometers for millimeter and submillimeter radiation measurements use a composite superconducting structure with metal film absorber on a low-heat-capacity, dielectric substrate. This type of bolometer has lower sensitivity and has limited applications.

A high-temperature transition-edge bolometer made from YBCO film ($T_c = 91$ K) on strontium titanate substrate has demonstrated an NEP as low as 7×10^{-12} W/$\sqrt{Hz}$, which is impossible to meet by a conventional room-temperature IR detector for measurement of radiation with wavelengths greater than 13 µm.

8.2.1 Superconducting Hot Electron Bolometer

The sensitivity of the Nb-based SIS quasi-optical detectors, which are widely used in astronomical sensors, deteriorates rapidly at frequencies greater than the superconducting gap frequency, which lies between 700 and 750 GHz for high-quality Nb films. An SIS junction based on niobium nitride (NbN) can solve this problem, because it has a gap frequency in the vicinity of 1400 GHz. Superconducting hot electron transition-edge (SHETE) bolometers can be alternatives to SIS-junction bomometers at about 1000 GHz. SHETE devices rely only on electron heating, and the detector response is not limited by the superconducting gap frequency. However, there is an upper limit for the if frequency due to finite time required to cool the hot electron detector to a specified cryogenic temperature. The response time is too short to allow for useful if frequency operation. This bolometer is a short and narrow microbridge structure with large normal-gold films. The short bridge ensures that the thermal conductance associated with out-diffusion of the hot electrons into the gold will dominate over that of the electrons [4].

The if frequency rolloff is dependent on the thermal time response of the detector. A response time of 55 ps has been demonstrated, which corresponds to if frequency rolloff of 2900 MHz, which can be increased to 4000 MHz with a 0.2-µm-long microbridge configuration. The if frequency rolloff of 2.9 GHz at 533-GHz operation is a significant improvement over that of the competing photon, cryogenically cooled bolometer. Diffusion-cooled bolometers [4] are expected to dominate the market for very high millimeter-wave (MM-wave) frequency (> 1000 GHz) measurements. They offer low noise and wide if bandwidth at very high MM-wave frequencies.

8.2.2 Superconducting Hot Electron Microbolometer

Optically coupled, cryogenically cooled, bolometer detectors are widely used for astronomical photometry at MM-wave and sub-MM-wave operations to achieve high sensitivity. However, superconducting antenna-coupled microbolometers have distinct advantages in terms of heat capacity and optical efficiency [5]. A hot electron microbolometer (Figure 8.8) uses a thin film of

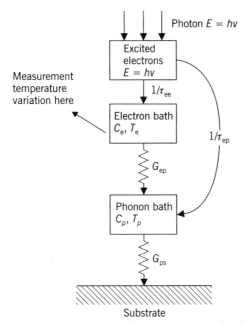

FIGURE 8.8 Block diagram of a hot electron microbolometer [5].

copper connected between the antenna terminals (not shown). The rf current from the superconducting antenna is dissipated into the resistive copper strips, resulting in a temperature rise in the electrodes and voltage across the aluminum superconductor–insulator–normal (SIN) metal tunnel junction deposited on the copper strip. Desired direct relaxation of the energy of the excited electrons to photons can be faster than relaxation to photons by way of hot electron bath.

Both the electron–electron and electron–photon relaxation rates govern the operation of the microbolometer with temperature as low as 300 millikelvins (mK). Hot electron microbolometers have demonstrated satisfactory operations in the 750- to 970-GHz range with absorption power as low as 140 pW, photon-to-signal temperature differential of 33 mK, electron-to-signal temperature differential of 36 mK, photon noise less than 1.6×10^{-31} W^2/Hz, and junction noise less than 2.0×10^{-32} W^2/Hz. Hot electron microbolometers operate at 4.2 K and have potential applications in both ground-based and spaceborne astronomical photometry systems.

8.3 SEMICONDUCTOR LASERS

Improvement in performance of cryogenically cooled semiconductor lasers, including single-quantum-well (SQW) lasers, double-quantum-well (DQW) lasers, multiple-quantum-well (MQW) lasers, and strained-layer QW lasers is discussed in this section. The semiconductor laser systems operating over the 1.8- to 2.1-μm range have potential applications in optical communication, eye surgery, spectral analysis, and laser pumping source. Solid-state lasers operating in the midinfrared region (2–5 μm) provide stable sources for laser radars, remote sensors, pollution-monitoring systems, and molecular spectroscopes. It is possible to generate higher optical powers in the mid-IR region using co-doped and undoped rare-earth laser crystals. A cryogenically cooled SQW laser diode (Figure 8.9) offers significant improvement in conversion efficiency, threshold current, and power output level. Similarly, other semiconductor laser diodes when cooled down to appropriate cryogenic temperatures will offer much improved performance over room-temperature (300 K) operation.

8.3.1 Cryogenically Cooled InGaAsP/InP Semiconductor Lasers

Significant performance improvement is expected from the cryogenically cooled InGaAsP/InP SQW, DQW, and MQW semiconductor lasers operating around 2 μm. Lasers operating around this wavelength are most attractive for medical uses, eye-safe illuminators, solid-state pumping lasers (SSPLs), and military applications. Improvement in threshold level and output power for an SQW semiconductor laser as a function of operating temperature is shown in Figure 8.10. Higher powers are possible with cryogenically cooled DQW laser diodes. An InGaAsP/InP–DQW-heterostructure epitaxial laser diode offers more than twice the power at 77 K compared to that at 300 K. This laser offers total external efficiency better than 55%, internal quantum efficiency better than 73% and overall differential quantum efficiency better than

p^+ InGaAs	2500 Å
p InP	1.5 μm
InP	1500 Å
InGaAsP (1.22 μm)	2200 Å
▬▬▬▬▬▬▬▬▬▬	Quantum well
InGaAsP (1.22 μm)	2200 Å
n InP	8000 Å
n^+ InP substrate	

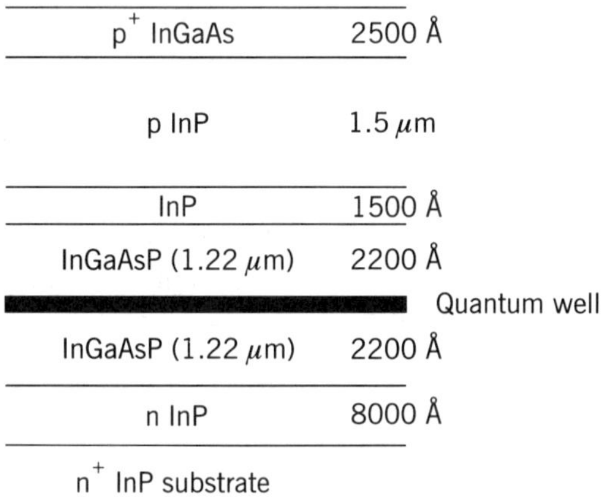

FIGURE 8.9 Structural layers of a cryogenically cooled, single-quantum-well (SQW) diode laser.

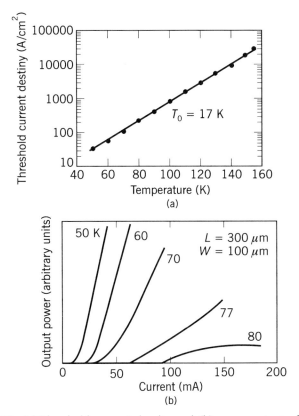

FIGURE 8.10 (a) Threshold current density and (b) power output of a cryogenically cooled semiconductor laser [6].

44% even at 300 K operation. An improvement of better than 30% in these parameters is expected at 77 K.

8.3.2 Cryogenically Cooled InAsSb/AlAsSb and InAsSb/InAsSbP Lasers

Both these laser diodes when fabricated on high performance GaAs substrate offer improved performance over the 2- to 4- µm range, because their structures provide the best combination of optical confinement, electrical confinement, and wide wavelength coverage (1.7 to 4.4 µm with lattice matched layers and longer wavelengths using compressed, strained quantum wells). An InAsSb/AlAsSb-double-heterostructure (DH) laser diode [6] demonstrated a pulsed threshold current density at 260 A/cm^2, differential quantum efficiency of 70%, and single-ended cw power output greater than 600 mW at 77 K operating temperature, which represents significant improvements in these parameters compared to those at 300 K. When operating at 4 µm this device exhibits cw operations at temperatures up to 80 K and pulsed operations up to 150 K. The lowest threshold current density occurs around 50 K for these semiconductor lasers (Figure 8.10).

8.3.3 Cryogenically Cooled Quantum-Cascade Lasers (QCL)

Semiconductor QCLs offer moderate power levels in the 5- to 8-µm range at cryogenic operations. A 5.2 µm QCL demonstrated a peak pulsed optical power of 200 mW, an average power of 7 mW, and slope efficiency of 106 mW/A at 300 K. The same diode is expected to yield more than 30% improvement at 77-K operation. However, a 8.5 µm QCL diode yielded a peak power of 1 mW at 300 K, 3 mW at 140 K, and 10 mW at 77 K, besides remarkable improvement in current density at 77 K. These data indicate that the cryogenically cooled QCL laser diodes offer improved performance at longer wavelengths. Much higher power levels are possible from QCLs compared to DQW-laser diodes when operated well below 77 K.

8.3.4 Cryogenically Cooled Diode Laser With Low Threshold Current

Russian scientists [7] have demonstrated improved performance of a cryogenically cooled diode laser with very low lasing current, thereby yielding significant improvement in optical efficiency and device reliability. Lasing was obtained over the 2.7- to 3.9-µm range in a DH-InAsSb/InAsSbP laser diode. The laser exhibited threshold current as low as 40 mA in a cw regime [7]. Laser operating wavelengths and lasing threshold currents at 80 K are shown in Figure 8.11. This particular laser offers tuning capability as a function of either lasing current or operating temperature. The characteristic temperature (T_o) is about 25–30 K for a wideband laser operation and about 20–25 K for a narrowband laser operation. The total differential quantum efficiency (TDQE) for both facets is 14 to 17% for the wide laser (2.9–3.5 µm), which is very impressive for a laser operating at 82 K. Both power output and TDQE can probably be further improved if the temperature is reduced below 50 K. The lasing modes shift to the short-wavelength side as the drive current increases. The short-wavelength modes dominate at higher drive currents at an operating temperature of 80 K. Nevertheless, the rate of shift, the direction of shift, and the drive current level will be different if the operating temperature is reduced to, say, 50 K.

8.4 SOLID-STATE DIODE PUMP LASER

Solid-state diode pump (SSDP) lasers are capable of generating higher optical power levels at longer wavelengths, using two different rare-earth non-co-doped crystals. Critical elements of an SSDP laser using such crystals are shown in Figure 8.12. One can see that a diode-pumped Tm:YLF laser can pump a Ho:YLF laser emitting at 3.9 µm, with appropriate optics maintained

8.4 SOLID-STATE DIODE PUMP LASER   203

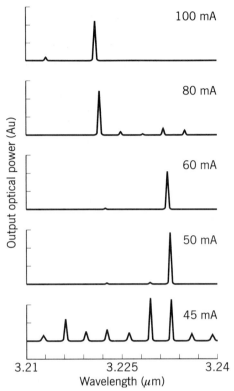

FIGURE 8.11 Lasing spectra for the semiconductor InAsSb/InAsSbP laser diode as a function of drive current [7]. Au = arbritary unit.

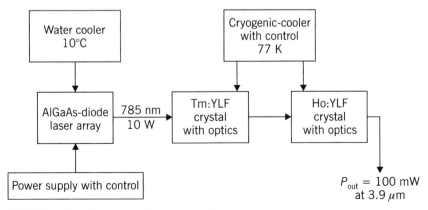

FIGURE 8.12 Cryogenically cooled solid-state pump laser design configuration.

at right cryogenic temperatures. Tm and Ho lasers operate efficiently only at crogenic temperatures of 77 K or lower. All rare-earth ions in Tm and Ho are commonly co-doped in a crystal. Thus, one can use a Tm:YLF laser to pump a Ho:YLF laser or use a YLF crystal with both Tm and Ho doping in the same crystal. A diode laser array of AlGaAs at 785 nm can be used to excite the Tm from where the laser energy is transferred directly to Ho, which will then lase at 2.1 µm. The AlGaAs diode array will need cooling at 10°C or lower to achieve higher conversion efficiency. However, both the Tm:YLF and Ho:YLF laser crystals need cryogenic cooling at 77 K or lower to obtain higher optical power and efficiency, which can be further improved if the operating temperature is reduced to 40 K or lower.

8.5 IR IMAGING CAMERA

An IR imaging camera capable of sensing 8.5-µm light has potential applications in security; surveillance; navigation; flight control; early warning; detection of electrical/or mechanical failures in nuclear power stations, transformers, circuit breakers, and switches; and monitoring of structural defects. The IR image sensor operation is based on a 256-pixel × 256-pixel focal planar array (FPA) of a quantum-well-infrared photodetector (QWIRP). When a quantum well is extremely deep and narrow, its energy states are quantized. The two energy states include a ground state near the well bottom and a first excited state near the well. A photon striking the well excites a ground-state electron to the first excited state, where externally applied voltage sweep it out, producing a photocurrent. Only the photons having energies corresponding to the energy separating the two states are absorbed by the detector, indicating a sharp absorption in the spectrum.

The QWIRP performance depends on minimizing the dark current that flows through a biased detector when no photons are impinging on it. The magnitude of the dark current can be controlled by the operating temperature of the detector in the QWIRP. At operating temperatures less than 30 K, sequential tunneling of ground-state electrons from well to well dominates the dark current, which can be partially reduced by widening the barriers. At temperatures above 50 K, thermionic emission dominates. Dropping the first excited state to well top causes the dark current to drop by a factor of three at an operating temperature of 70 K. The dark current can be further reduced by lowering the temperature to around 40 K.

In addition to dark current, a key factor in the performance of the QWIRP-FPA is the light-coupling mechanism. The GaAs-quantum wells are separated by AlGaAs barriers. The light striking the quantum-well layer is generally not absorbed completely, unless optical mirrors are used to bend the light transverse wave. Absorption requires bending the incident light inside the detector.

The responsivity (R^*) and detectivity (D^*) are two figures of merit for comparing the performance of photodetectors. The higher the responsivity the greater the sensitivity of the detector, which is possible only at cryogenic operations. Detectivity is the signal-to-noise (S/N) ratio normalized to unit area and unit bandwidth. The main noise source in a QWIRP sensor is the shot noise caused by the dark current, which can be reduced at cryogenic temperatures of 77 K or lower. Thus, the performance of a QWIRP sensor can be significantly improved by cooling the FPA assembly to an appropriate cryogenic temperature, which will reduce the dark current exponentially as a function of operating temperature. The detectivity for a 0.9-μm QWIRP is better than 10^{11} cm $\sqrt{Hz}$ at 70 K and rises exponentially with cryogenic cooling. This value of detectivity is more than sufficient to meet the demand of large two-dimensional FPAs with 256×256 pixels or more at long wavelengths.

8.6 STARING IR CAMERA

A staring IR camera offers several advantages: extended camera life, fail-safe imaging, elimination of mechanical components, highest reliability due to absence of wear and tear parts, and high-quality imaging. The IR scanning camera uses an array of tens of thousands of cryogenically cooled IR detectors at 77 K or lower. The entire array of detectors stares at the scene and provides an image with update rate of 30 to 60 times/s, thereby providing a dwell time as long as 16,000 times that of a conventional IR camera. This particular camera offers a high-quality image across the field of view (FOV). An IR camera with 12-bit digital storage capacity provides images with 4096 distinct color levels. The quality of the image can be further improved if the detector array is cooled well below 50 K.

8.7 MONOLITHIC CHARGE TRANSFER DEVICES

Charge-injection device (CID) and charge-coupled device (CCD) technologies are based on the technology used for charge-transfer device (CTD) development. Both CIDs and CCDs are used in the development of monolithic Hg:Cd:Te IR imaging sensors optimized for the 3- to 5-μm operating range. These devices perform some signal processing before the signal is transferred from the cryogenically cooled, Hg:Cd:Te monolithic detector arrays to a silicon signal-processing chip. Both the time-delay integration (TDI) and multiplexing (MPX) are performed by the CCD devices. CIDs and CCDs utilize the same Hg:Cd:Te compound and metal–insulator–semiconductor (MIS) fabrication technology. A 32-stage CCD when cooled down to 77 K is capable of operating as an IR sensor with TDI enhancement of sensitivity and background-limited detection capability. Cryogenically cooled, monolithic Hg:Cd:Te CCD and CID arrays offer much improved IR images that are not possible otherwise.

8.8 IR LINE SCANNER

IR line scanners are widely used by reconnaissance aircraft to locate targets of military importance. The IR line scanner is installed at the bottom of the fuselage centerline to provide IR images of the targets on both sides of the aircraft. The IR line scanner uses a linear array of cryogenically cooled Hg:Cd:Te detectors at 77 K. The line scanner sweeps the IR image perpendicularly across the detector array twice using a scanning motor to provide 1.7×10^7 pixels per frame.

8.9 FORWARD-LOOKING INFRARED (FLIR) SYSTEM

FLIRs are widely used in night vision and surveillance systems for military and civilian applications. The performance of a FLIR depends on the number of detectors employed, the sensitivity of the detectors, and operating temperature. A Hg:Cd:Te detector array comprising of seven cryogenically cooled (77-K) detector elements offers reasonably good performance over the 8- to 14-µm range.

A 200-detector element array comprising four small detector array chips each having 50 detectors has demonstrated superior performance, as indicated in Table 8.3. The last column indicates the overall detection capability of the entire array comprising 200 detector elements. Increased detectivity (D^*) can be obtained by increasing the integration time for which one sensor element integrates a signal from one image pixel and by operating the detector array at a lower cryogenic temperature. In a scanned mode, higher integration time must be obtained by the TDI of the signal from several detector elements. The image pixel is sensed as it passes sequentially along a row of the detectors operated at 77 K and all the signals from the same pixel are added together to obtain high-quality IR image of the target over a wide spectral range. The maximum detectivity of the entire array, which is about $3 \times 10^{10}\,\text{cm}/\sqrt{\text{Hz}}\,\text{W}^{-1}$ at 300 K, improved to $4.6 \times 10^{10}\,\text{cm}/\sqrt{\text{Hz}}\,\text{W}^{-1}$ at 4.2-K operating temperature.

TABLE 8.3 Detectivity of Each Chip and Whole Array at Various Tempertures

Temperature (K)	Detectivity ($\text{cm}/\sqrt{\text{Hz}}\,\text{W}^{-1}$)				
	1st Chip	2nd Chip	3rd Chip	4th Chip	Array
77	4.0×10^{10}	3.3×10^{10}	2.7×10^{10}	2.9×10^{10}	3.2×10^{10}
10	6.0×10^{10}	4.0×10^{10}	3.3×10^{10}	3.6×10^{10}	3.8×10^{10}
4.2	7.2×10^{10}	5.3×10^{10}	4.8×10^{10}	5.0×10^{10}	4.6×10^{10}

8.10 IR SENSOR USING CRYOGENICALLY COOLED MULTICHANNEL CCDS

Low-temperature operation of silicon CCDs is necessary if the CCD devices are to be used in an IR sensor. The operating temperature of IR CCD is a function of a number of factors, namely, type of CCD element, the wavelength region of interest, the IR detector material, and the sensitivity requirement. Both the bulk and surface channel CCDs are of great interest for application in IR sensors. The epitaxial bulk or peristaltic-channel CCD (PCCD) (Figure 8.13) is of greatest interest, because of its high-frequency capability is well suited for large IR focal planar array applications, where a high data rate is required for readout of all detectors in one frame. The transfer inefficiency per cell of a PCCD is its figure of merit, which decreases over the temperature range of 77 to 30 K. The image pattern is reversed, with the transfer inefficiency slowly increasing with the decreasing temperature, and reaching a saturation level of 0.2 at 4.8 K. The transfer inefiency per cell of a surface channel CCD (SCCD) has a minimum value of 0.004 at 77 K with saturation occurring at 5 K, similar to PCCD performance. These devices are best suited for military applications, where high resolution from IR cameras is the principal requirement.

8.11 SIS QUASI-PARTICLE RECEIVER MIXER

The SIS quasi-particle (SISQP) mixer is an important element of a superconducting infrared receiver. An SISQP heterodyne mixer is important, particularly at MM-wave frequencies. Very low capacitance and very low current density junctions, which are possible only at low cryogenic temperatures, are essential for high-T_c quasi-particle devices to operate in the terahertz range. The noise temperature of the receiver that uses cryogenically cooled Schottky diode mixers is less than 10 times that of the SIS receiver at the W-band (75–110 GHz), because of the energy gap limitation to the operating frequency of the SIS mixer. This limitation has been eliminated by the use of a high-T_c superconductor such as NbN, for which the T_c is 15 K, compared to 9 for Nb and 7 K for Pb. The requirement of small current flow at voltages below ($2\Delta/e$) sets an upper limit of about $T_c/2$ (2Δ is the energy gap, e is electronic charge, and T_c is the critical temperature of the superconductor) on the operating temperature that can be used for SIS mixer operation.

Direct response of a SISQP mixer (Figure 8.14) is important over the 3.1- to 4.2-K temperature range for the mixers operating over 500 to 700 GHz. The direct response is much better than that for Schottky diode mixers at these frequencies. These mixers are best suited for coherent receivers for MM-wave telescopes and interferometers at observatories. They are also used in portable coherent receivers for measurements of molecular lines at submillimeter frequencies on mountaintop and airborne telescopes. Thin-film SIS tunnel

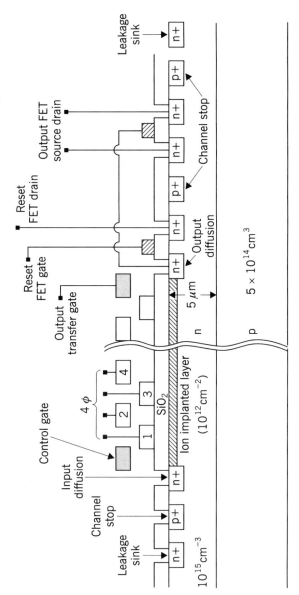

FIGURE 8.13 Cross section of PCCD in the direction of charge propagation.

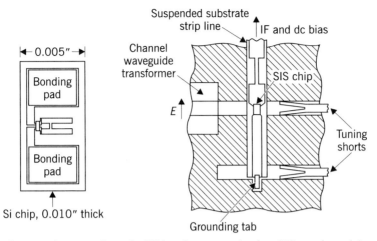

FIGURE 8.14 Cross section of a W-band superconducting SIS quasi-particle mixer.

junctions are compatible with lithographic superconducting receivers components such as planar antennas (Figure 8.15), transmission lines, and filters.

8.12 QUASI-OPTICAL SIS (QOSIS) RECEIVER

Cryogenically cooled QOSIS receivers offer optimum performance over extremely broad bands from 100 to 800 GHz, because the gap frequency of an NbN junction is around 1500 GHz. However, these receivers have some drawbacks, such as higher shot noise, quantum noise, and Josephson effect noise. The first two noise components can be reduced by operating at less than 50% of the critical temperature of the NbN junction, while the last component, which produces mixer instability, can be eliminated by operating the mixer above the threshold voltage.

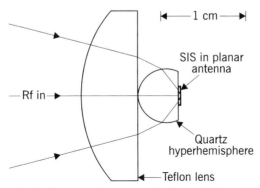

FIGURE 8.15 Block diagram of a cryogenically cooled quasi-optical SIS mixer integrated with a lens antenna.

8.13 OPTICAL MODULATOR

Superconducting electrodes can be used in a resonant-type lithium-niobate (LiNbO$_3$) optical modulator [8]. Preliminary studies indicate that superconducting electrodes can significantly improve optical modulator in term of modulation depth and power. Driving power for the optical modulator [8] with superconducting electrodes (Figure 8.16) can be significantly reduced at liquid helium temperature. Modulation power and depth for cryogenically cooled gold and lead electrodes at 14.8 GHz are shown in Figure 8.17. The modulation power [8] requirement is 140 mW for the Pb-alloy electrode compared to 40 mW for the gold electrode at 4.2 K. Sharp modulation depths can be observed at lower cryogenic temperatures with both types of electrodes. The reflection coefficient increases by about 12 dB when the operating temperature is increased to 10 K from the normal operating temperature of 4.2 K.

8.14 IR COMMUNICATION TRANSMITTER USING CRYOGENICALLY COOLED LIGHT-EMITTING DIODES

Light-emitting diodes (LEDs) offer a high data rate for the carrier used in the IR transmitter system. The IR transmitter uses a series of cryogenically cooled GaAs LEDs. The spontaneous emission of the diode is the result of energy released when the electrons and holes combine in the junction of a forward-biased semiconductor diode. The radiation is proportional to the current through the diode and increases as the current increases until it is dissipation-limited by the junction temperature. GaAs LEDs are fully developed and are widely used for IR applications because of minimum cost and abundant supply.

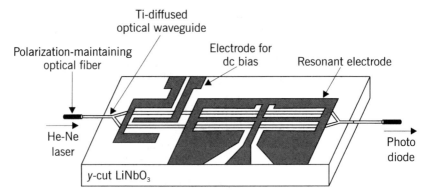

FIGURE 8.16 Schematic diagram of a cryogenically cooled LiNbO$_3$ optical modulator with a resonant electrode [8].

8.14 IR COMMUNICATION TRANSMITTER USING CCLED

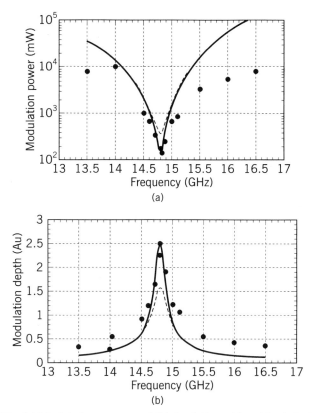

FIGURE 8.17 (a) Modulation power and (b) depth for gold and lead electrodes at 4.2 K [8]: •, Pb, experimental; —, Pb, theoretical; ---, Au, theoretical.

Power output, efficiency, and reliability of the LEDs at 300 and 77 K are summarized in Table 8.4. One can see remarkable improvements in these parameters at 77 K. Further improvement is possible, if the operating temperature is reduced below 50 K.

TABLE 8.4 Characteristics of GaAs LEDs as a Function of Temperature

LED parameter	Temperature (K)	Values
Peak power output (mw)	300	1–50
Average power output (mw)	300/77	3/6
Power efficiency (%)	300/77	5/25
Reliability (MTBF) (h)	300	M^a
Reliability (MTBF) (h)	77	$> 3M^a$

[a] This is an estimated value in terms of relative mean time between failures (MTBF) at 300-K operating temperature.

IR communication systems using LED elements are capable of handling high data rates, but they are limited to short distances. IR systems have potential applications in ship-to-ship communication. Both the operating range and reliability can be further improved if the LEDs can be operated at 30 K or less.

8.15 SUPERCONDCUTING SCANNING SQUID MICROSCOPE

A superconducting scanning SQUID microscope (SSSM) design is based on its critical element SQUID, which must operate at a specified cryogenic temperature for optimum resolution. This cryogenically cooled instrument can image magnetic fields at the surface of a sample under observation with impressive sensitivity and resolution. It can sense magnetic fields of less than 100 picoteslas, currents of few nanoamperes, and magnetic moments of few hundred Bohr magnetrons. Furthermore, the SSSM has demonstrated remarkable spatial resolution down to 4 μm and ultrahigh sensitivity from dc to a gigahertz at 77 K, which can be further improved if the operating temperature is reduced to 4.2 K. The SSSM must be housed in a cryogenic enclosure for optimum performance. Spatial resolution and sensitivity can be optimized by selecting suitable pickup coil size, superconducting material, and operating temperature.

SUMMARY

Cryogenically cooled detectors, bolometers, laser diodes, mixer diodes, and CCD devices will realize significant improvement in terms of quantum efficiency, sensitivity, reliability, and detectivity. These devices are used in SIS receivers, laser radars, EO sensors, FLIRs, IRSTs, optical modulators, IR communication systems, scanning microscopes, IR receivers, multispectra surveillance systems, IR cameras, satellite imaging sensors requiring high optical resolution, and focal planar array-based imaging systems.

REFERENCES

1. W. L. Wolfe and G. J. Zissis. *The Infrared Handbook*. Environmental Research Institute of Michigan, 1978, Chapter 11, pp. 66–86.
2. A. Jager et al. The SUPFET: a new phtodetector with ultrathin YBaCuO/PrPaCuO multilayer channel. *IEEE Trans. Appl. Superconductivity* 5:2865–2868 (1995).
3. V. Yu et al. Fourier transform spectrum studies (300–1000 GHz) of Nb-based quasi-optical SIS detector. *IEEE Trans. Appl. Superconductivity* 5:3445–3450 (1995).
4. A. Skalare et al. A heterodyne receiver at 533 GHz using a diffusion-cooled–hot-electron mixer. *IEEE Trans. Appl. Superconductivity* 5:2236–2239 (1995).
5. A. Tang et al. Quantum effects in the hot-electron microbolometer. *IEEE Trans. Appl. Superconductivity* 5:2599–2602 (1995).

6. S. J. Eglash et al. InAsSb/AlAsSb double-heterostructure-diode lasers emitting at 4 microns. *Appl. Phys. Lett.* 64:833–835 (1994).
7. A. N. Baranov et al. 2.7–3.9 micron InGaSb(P)/InAsSbP low-threshold diode lasers. *Appl. Phys. Lett.* 64:2480–2481 (1994).
8. K. Yosida et al. Application of superconducting electrodes to resonant-type $LiNbO_3$ optical modulator. *IEEE Trans. Appl Superconductivity* 5:3183–3185 (1995).

CHAPTER NINE

Applications of LTSC and HTSC Technology to Medical Diagnostic Equipment

This chapter describes the clinical diagnostic devices and equipment using low- and high-temperature superconducting technologies. For more than 20 years superconducting quantum interference device (SQUID) magnetometers have been used to measure biomagnetic signals for heart, brain, liver, nerves, eyes, and other organs of the human body. Advances in digital SQUID and high-temperature (high-T_c) superconducing technologies will play a significant role in optimizing SQUID systems so that they can provide accurate and reliable biomagnetic measurements at minimum cost. Emphasis is placed on the clinical aspects of SQUIDs using LTSC technology because they have greater sensitivity at lower cryogenic temperatures. SQUID magnetometers have demonstrated unique diagnostic capabilities in medical fields where there are no reliable alternative techniques. Conceptual and block diagrams of a typical SQUID magnetometer that is widely used in biomedical applications are shown in Figure 9.1.

Bioelectric signals are used as a diagnostic tool in medical fields; an example is an electrocardiogram (ECG). The analogous magnetic measurements are known as magnetocardiogram (MAG) and magnetoencephalogram (MEG). The magnetic measurement has no analog. Magnetic fields from active electrical sources in the body can be measured passively and external to the body by placing the superconducting SQUID magnetometer in close proximity to the body surface. SQUID-based gradiometers using superconducting detection coils allow biomagnetic measurements in unshielded environments at sensitivities well below $20\,\text{fT}/\sqrt{\text{Hz}}$. A SQUID magnetometer measures the functional activity of human body organs, rather than providing structural information like magnetic resonance imaging and computer tomography.

9.1 PERFORMANCE CAPABILITIES AND DESIGN ASPECTS OF SQUIDs

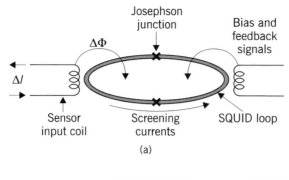

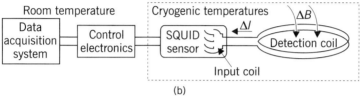

FIGURE 9.1 Block diagram of (a) a SQUID sensor and (b) a SQUID magnetometer.

By mapping neural activity it is possible to locate the electrical regions of interest in the brain within few millimeters, which can provide reliable detection of various neurological disorders, including epileptic seizures. Applications of various SQUID devices in ECG, MAG, and MEG are described with emphasis on temperature-dependence sensitivity, reliability, bandwidth, and affordability.

9.1 PERFORMANCE CAPABILITIES AND DESIGN ASPECTS OF SQUIDs

A SQUID device consists of one or more Josephson junctions (JJs) and associated electronics. A JJ is a weak link between the two superconductors that can support a current below its critical current, I_c. As illustrated in Figure 9.1, an RF-SQUID uses a single JJ that is connected to a superconducting loop and an rf biasing current circuit inductively coupled to the SQUID device to measure impedance (Z). In contrast, a DC-SQUID uses a superconducting loop containing a pair of JJs with bias current directly applied to the SQUID element to measure the loop inductance. Regardless of SQUID classification, the JJs allow the impedance of the SQUID loop to be a periodic function of the magnetic flux linking SQUID loop, so that a modulation signal can be applied. The bias current can be used with a lock-in detector to measure the loop impedance and to linearize the voltage-to-flux relationship. In brief, a SQUID functions as a flux-to-voltage converter with unique energy sensitivity, which can play a significant role in the design and development of a sensitive

216 LTSC AND HTSC TECHNOLOGY IN MEDICAL DIAGNOSTIC EQUIPMENT

SQUID magnetometer capable of measuring weak biomagnetic signals. Sensitivity, bandwidth, and noise level are the most critical performance parameters of both types of SQUIDs.

9.2 DERIVATION OF CRITICAL PERFORMANCE PARAMETERS OF A SQUID

Because of the superconducting nature of the JJ effect, it is necessary to maintain the SQUID temperature well below the critical temperature of the superconducting material used in the fabrication of the device. High sensitivity and low noise are the principal requirements for a SQUID device, regardless of its use.

9.2.1 Sensitivity

The sensitivity of a SQUID is defined as the ratio of flux density in the pickup coil to the amount of flux linkage. This means that the sensitivity can be defined as

$$\frac{B}{\phi} = \frac{1\,\text{nT}}{\phi_0} \tag{9.1}$$

where B is the flux density, ϕ is the flux linking the coil, ϕ_0 is the flux quantum ($2.06 \times 10^{15}\,\text{v}\cdot\text{s}$ or weber), and 1 nT (nanotesla) is the minimum intensity of the magnetic field equal to $1/10^{15}$ t.

9.2.2 Flux Density Noise Level ($\sqrt{S_\phi}$)

The flux density noise level is defined as

$$\sqrt{S_\phi} = 1\,\text{fT}/\sqrt{\text{Hz}} \tag{9.2}$$

where ft stands for femtotesla.

9.2.3 Flux Noise Level (S_ϕ)

The corresponding flux noise level can be written as

$$\sqrt{S_\phi} = \mu\phi_0/\sqrt{\text{Hz}} \tag{9.3}$$

where μ is the permeability of the medium.

9.2.4 Bandwidth and Slew Rate

Bandwidth and slew rate are the critical operational requirements besides the sensitivity for biomagnetic measurements. The SQUID bandwidth is defined as the frequency bandwidth over which a specific value of flux density noise level is desired, and is expressed in terms of corner frequency (f). A corner frequency close to 1 Hz is most desirable for biomagnetic measurements. According to published data [1], an upper flux density noise level limit for a sensitive biomagnetic system with real diagnostic value is about $10\,fT/\sqrt{Hz}$ for a corner frequency of 10 Hz. A system bandwidth of about 2 kHz is most desirable for measurements of evoked signals with adequate slew rate, which is determined by maximum peak amplitude of signal.

Clinical researchers indicate that SQUID systems with a bandwidth of 10 kHz and a slew rate of $10^4 \Phi_0/s$ are more than adequate for all biomagnetic sources. In addition to signal dynamics, electrical environmental disturbance up to 50–60 Hz due to power line interference must be considered, which has a typical amplitude of 10 pT corresponding to a slew rate of $3\Phi_0/s$ inside a magnetic shielded room, where the constant Φ_0 is equal to $2.06/10^{15}$ Weber. Without a shielded room, the power line interference is about 300 nT, corresponding to a slew rate of $10^6 \Phi_0/s$, which will require a very fast SQUID system with higher slew rates.

To obtain reliable MEC and MAG measurements with low-temperature SQUIDs, magnetically shielded rooms (MSRs) are necessary because of high S/N ratio requirements. High-performance MSRs offer S/N improvement of 50 dB at 0.2 Hz, 56.5 dB at 1 Hz, and 60 dB at 12 Hz, which is most desirable for reliable biomagnetic measurements [2].

9.2.5 Power Spectral Density and Energy Efficiency

The power spectral density $S_\phi(f)$ and the energy efficiency or energy resolution per unit bandwidth $\varepsilon(f)$ are other critical parameters, which are functions of pickup coil and input coil parameters and operating temperature.

The energy efficiency is defined as follows:

$$\varepsilon(f) = \frac{S_\phi(f)}{2L} \qquad (9.4)$$

where L is the inductance of the input coil. The SQUID can act as an amplifier and the gain of a SQUID amplifier (G_a) is defined as the ratio of flux quantum to flux quantum in the signal coil. This means that G_a can be written as

$$G_a = \frac{\phi_0}{\phi_s} \qquad (9.5)$$

where the subscript s stands for signal. This parameter is a function of pickup coil inductance and its area and signal coil inductance and its area.

In summary, SQUID sensitivity is the most important performance parameter of a superconducting SQUID magnetometer and is directly proportional to the operating temperature. Preliminary calculations indicate that a SQUID magnetometer operating at 77 K would be nearly 20 times noisier than that one operating at 4.2 K. Another critical performance parameter is the white flux noise $\sqrt{S_\phi}$, which is also temperature-dependent. Its typical value for a DC-SQUID at 4.2-K operating temperature is about 2×10^{-6} J/Hz, which is equal to $400h$ ($h = 10^{-34}$ J·s. At higher cryogenic temperatures, the magnitude of white flux noise is significantly high. However, the use of a modulation feedback coil (Figure 9.2) not only eliminates the flux noise, but also improves the dynamic range and slew rates. Most commercially available SQUID sensors are made of niobium with T_c of 9 K, and the best noise performance and sensitivity are possible when the SQUID is operated at 50% of the critical temperature or lower. Typical

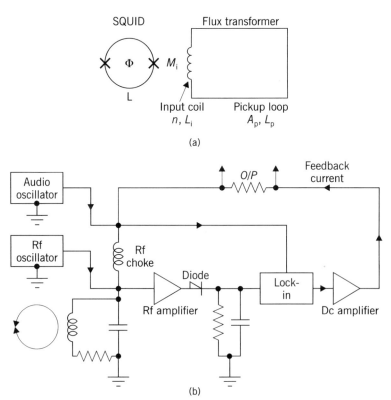

FIGURE 9.2 (a) SQUID elements; (b) circuit elements and critical components of an RF-SQUID.

values of frequency limits and magnetic field sensitivity requirements for various SQUID applications are illustrated in Figure 9.3. A SQUID can be used as an ammeter, voltmeter, ohmmeter, magnetometer, or gradiometer, but each will have its own sensitivity at 4.2 K.

9.3 CAPABILITIES OF VARIOUS SUPERCONDUCTING SQUID MAGNETOMETERS

SQUID magnetometers have been used worldwide to measure magnetic signals from the human heart, brain, liver, lungs, muscle, stomach, eyes, and intestines [3]. Biomedical scientists are now investigating the applications of advanced SQUID magnetometers using both the digital and high-temperature superconducting technologies. As stated earlier, a superconducting quantum interference device (SQUID) uses one or more Josephson junctions to support a supercurrent below a critical value of current (I). SQUIDs are classified into two categories: DC-SQUIDs and RF-SQUIDs. An RF-SQUID uses a single JJ that is connected to a superconducting loop. An rf current bias is inductively coupled to the SQUID device to measure the impedance. A DC-SQUID uses a superconducting loop with a pair of JJs and a dc current is applied to the SQUID to measure the loop inductance. Regardless of SQUID type, the special properties of the JJ cause the impedance of the SQUID loop coil to be a periodic function of the magnetic flux threading into the SQUID. A modula-

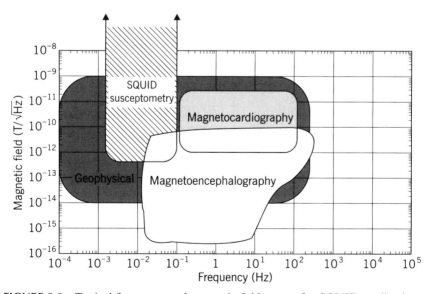

FIGURE 9.3 Typical frequency and magnetic field ranges for SQUID applications.

tion signal is applied to the bias current with a lock-in detector to measure the impedance and to linearize the voltage-to-flux relationship.

9.3.1 RF-SQUIDS

RF-SQUIDs are preferred over DC-SQUIDs where low cost and easy operation are the principal requirements, despite their lower sensitivities. Earlier RF-SQUID devices were made from thin-film tunnel junctions and torodial input and rf coils in a niobium (Nb) cavity maintained at or below 4.2 K. Specific details on the critical elements of an RF-SQUID are shown in Figure 9.4. This SQUID device has been used in a quasi-planar Josephson junction-type RF-SQUID magnetometer [4] using a superconducting thin-film microbridge configuration, which offers low sensitivity and good resolution over a wide temperature range of 3.2 to 5.2 K.

For stable operation of an RF-SQUID magnetometer, it is necessary to keep the critical current of the JJ to an optimum value over large temperature fluctuations. In the above-mentioned SQUID, a superconducting thin-film of RF-SQUID sensor element is inductively coupled to a Nb block with two holes. One hole of the sensor element is inductively coupled to an input coil, which offers a flux transfer loop with a magnetic field pickup coil. The other hole is inductively coupled to an rf LC-tank circuit. By inserting the SQUID

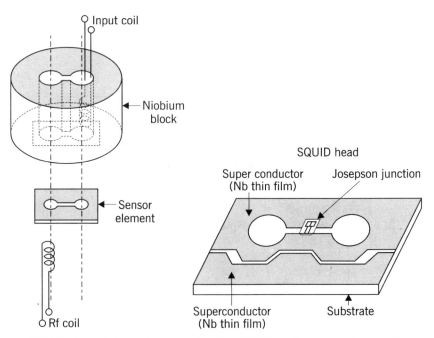

FIGURE 9.4 Critical elements of an RF-SQUID using planar design [4].

sensor and LC-tank circuit in a Dewar assembly and operating at optimum cryogenic temperature, one can improve the overall sensitivity of the RF-SQUID magnetometer.

Though less sensitive than a DC-SQUID, an RF-SQUID is adequate for most clinical applications except MAG and MEG applications. An RF-SQUID has a typical white noise energy of 5×10^{-29} J/Hz at a frequency of 0.1 Hz. SQUID sensitivity and magnetic field resolution can be improved by increasing the pickup coil size, SQUID coil inductance, and coupling coefficient between the input coil and pickup coil. However, in practice the size of the cryostat will impose the upper limit on the pickup coil size. By assuming a flux noise energy $\varepsilon(f)$ of 10^{-28} J/Hz and using coils with reasonable dimensions, one can achieve a magnetic field resolution of 5×10^{-15} T/$\sqrt{Hz}$ at an operating temperature of 4.2 K. Slightly higher sensitivity and resolution are possible if the operating temperature is reduced to 3.2 K.

Significant improvement in sensitivity and resolution is possible using superconducting flux transformers [5] (Figure 9.5). Improvement in these parameters is strictly due to significant reduction in white flux noise provided by flux transformers operating at temperatures of 4.2 K or lower.

9.3.1.1 Flux-Locked RF-SQUID Device Configuration Integration of superconducting flux transformers will lead to the development of first- and second-derivative gradiometers, which are best known for detection of weak neuromagnetic signals. A neuromagnetometer using a flux-locked RF-SQUID can detect the weak magnetic signals emanating from the human brain or eye, because such a gradiometer effectively discriminates against distant noise sources. Note an axial gradiometer using a flux-locked SQUID device actually senses the magnetic field rather than the gradient, because the distance from the signal source to the pickup loop is less than the baseline of the gradiometer. Typical magnetic field sensitivity of a well-designed pickup loop is about 10 fT/$\sqrt{Hz}$ at 4.2 K, which can be further improved if the operating temperature is reduced to 3.2 K.

A first-order gradiometer must be used in a magnetically shielded room (MSR) to reduce the ambient magnetic noise, while a second-order gradiometer, which has higher sensitivity, can be used in unshielded environments. A second-order gradiometer is best suited for measuring the weakest magnetic signal levels expected from the spinal cord, retina, and evoked cortical activities. Magnetic signal level ranges expected through cardiograms, gastrograms, encephalograms, and retinograms are summarized in Figure 9.6. While biological magnetic signal levels measured on various human organs are summarized in Figure 9.7. Superconducting SQUID-based gradiometers are discussed later in greater detail.

9.3.1.2 RF-SQUID Sensitivity as a Function of Number of Junctions and Current and Voltage Bias Loops The sensitivity of a 2-junction SQUID is higher than that of a single-junction SQUID, and the sensitivity of a 4-

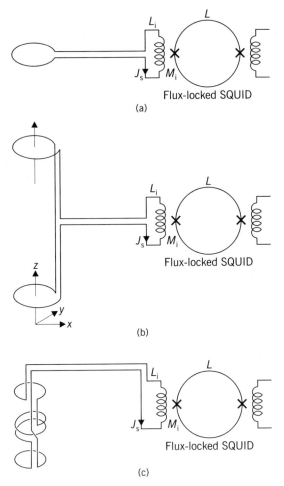

FIGURE 9.5 Potential flux transformer configurations for a SQUID [5].

junction SQUID is higher than that of a 2-junction SQUID. Both the cost and complexity increase with an increase in number of junctions.

SQUID magnetometers operate using both voltage and current bias loops. The bandwidth of a voltage-biased SQUID sensor is roughly three times narrower than that of a SQUID operating under current bias conditions. It is anticipated that the reduction in SQUID bandwidth when operated over the 3.2- to 4.2-K temperature range will significantly improve the sensitivity of the RF-SQUID sensor.

9.3.1.3 Three-Dimensional Vector SQUID Magnetometer
Professors at Osaka University [6] developed a 6-channel SQUID amplifier comprising of 6 RF-SQUID elements and 1 rf amplifier. Three channels were assigned to 3

9.3 CAPABILITIES OF VARIOUS SUPERCONDUCTING SQUID MAGNETOMETERS

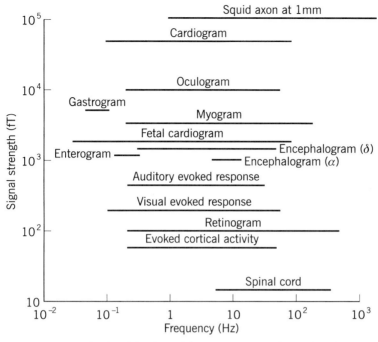

FIGURE 9.6 Magnetic signal levels and frequency ranges for various human organs.

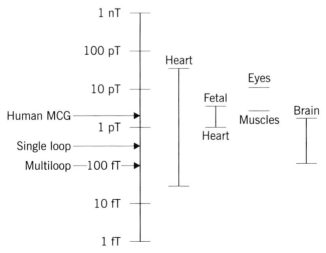

FIGURE 9.7 Measured biological magnetic field levels for various human organs. One tesla (1 T) = 10,000 G. [Measurements courtesy of S. J. Williamson et al. *J. Magn. Magn. Material* 22:129–209 (1981).]

orthogonal second-derivative flux transformers and the rest to 3 noise-canceling coils, as shown in Figure 9.8. The three-dimensional vector SQUID magnetometer (TDVSM) with 6 inputs is capable of simultaneously measuring the vector components of the heart magnetic fields using its 3 channels with automatic noise cancellation capability using the remaining 3 channels. All RF-SQUID elements must have same values of the critical currents needed for optimum system operation. This is accomplished using a Joules heating-type microbridge of variable thickness. The RF-SQUID elements are made from Nb thin-films and the critical current levels can be adjusted in liquid helium over 4 to 6 K. The 6-RF-SQUID elements are coupled to a LC-tank circuit and the rf carrier is modulated by the 6 input flux signals, which in turn modulate the subcarriers with separate frequencies in order to discriminate from each other. Weighted sums of the noise output components are subtracted from the second-derivative gradiometer output signals to cancel the noise components.

The TDVSM with automatic noise cancellation capability offers the most accurate information on the functional and structural aspects of the heart based on the vector magnetocardiogram measurements. This multichannel system using low-temperature superconductor technology offers highest sensitivity with minimum noise, thereby measuring the weakest magnetic field with high accuracy and reliability. It has potential applications in clinical research activities and diagnosis of heart and brain diseases.

9.3.1.4 RF-SQUIDs Using Digital Technology
Fujitsu (Japan) [7] developed a single-chip SQUID where the entire feedback loop is integrated on

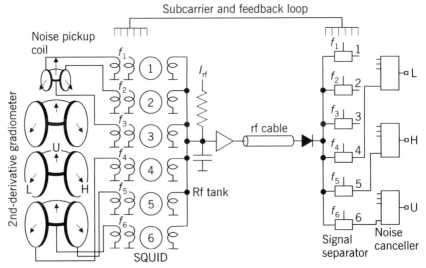

FIGURE 9.8 Three-dimensional vector SQUID magnetometer for improved MCG [6].

the SQUID chip, thereby combining both the digital and low-temperature (4.2-K) superconducting technologies. The device uses a 10-MHz ac bias to provide a series of voltage pulses at the output. An array of 8 such single-chip SQUIDs was developed to operate with a Josephson multiplex in order to reduce the number of cables between 4.2 K and room temperature (300 K). The magnetic flux noise and dynamic performance of this 8-channel SQUID array are considered more then adequate for biomagnetic measurements using the large gradiometer pickup coils necessary to connect the input coils. Fujitsu claims that both the noise and dynamic performance capabilities can be further improved by increasing the bias frequency into the 1-GHz range. Typical performance capabilities of the digital SQUID array are summarized in Table 9.1.

9.3.2 DC-SQUIDs

So far, performance capabilities and limitations of RF-SQUIDs have been discussed in the fields of clinical research and medical technology. This section describes the performance capabilities and limitations of DC-SQUIDs and their applications to SQUID-magnetometers and gradiometers for clinical research and biomagnetic studies. DC-SQUIDs can play a significant role as medical diagnostic tools in obtaining reliable and accurate clinical data involving MCGs and MEGs. DC-SQUID magnetometers have demonstrated better sensitivity than RF-SQUID magnetometers under similar operating conditions. DC-SQUIDs with feedback circuits are most attractive for biomagnetic multichannel systems because of their low noise, high slew rates, simple flux-locked loop electronics, and complete elimination of cross-talk. DC-SQUIDs are preferred where speed, accuracy, reliability, and superior neuromagnetic performance are the principal requirements. Energy sensitivity and noise levels are the most critical performance parameters. The parameters for various DC-SQUIDs operating at 4.2 K are shown in Figure 9.9.

9.3.2.1 First Generation of DC-SQUID Magnetometer
A simple schematic diagram for a DC-SQUID magnetometer includes a SQUID junction, tank circuit, input coil, modulation circuit, and lock-in detector. These

TABLE 9.1 Measured Performance Data on the Digital SQUID Array

SQUID Element Design Parameters	Noise Performance
SQUID coil inductance, $L = 35\,\text{pH}$	$16\,\text{kT}\sqrt{LC} = 4.1h$
SQUID coil capacitance, $C = 0.25\,\text{pF}$	Energy sensitivity $= 13{,}600h$
Input coil inductance, $L_i = 500\,\text{nH}$	$\sqrt{S_q} = 11.3\mu\phi_0/\sqrt{\text{Hz}}$

Note. $\sqrt{S_q}$ = system flux noise, h = Planck's constant ($10^{-34}\,\text{J}\cdot\text{s}$), μ = permeability constant ($4\pi \times 10^{-7}\,\text{H/m}$), and ϕ_0 = flux quantum (2.068×10^{-15}) Weber.

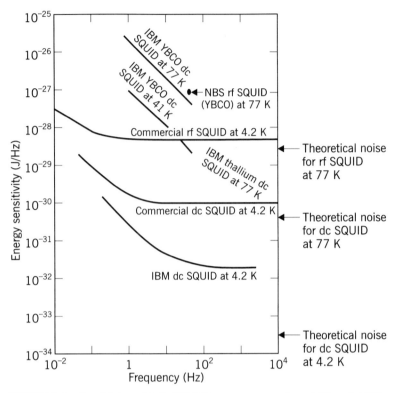

FIGURE 9.9 Sensitivity of DC-SQUID sensors operating at 77 and 4.2 K.

SQUIDs were originally known as hybrid DC-SQUIDs and were produced for many years by Biomagnetic Technologies (San Diego, CA) using a pair of tunnel junctions in a cryogenically cooled toroidal coupling cavity.

Implementation of additional positive feedback (APF) and bias control feedback (BCF) (Figure 9.10) have reduced the preamplifier flux noise contribution and canceled the preamplifier current noise. A SQUID with an APF circuit can be operated in either the current-biased or the voltage-biased mode. The maximum bandwidth with voltage bias is about three times narrower than that of the current bias operation. However, the voltage bias offers slew rates four times higher than the current bias.

9.3.2.2 DC-SQUIDs With Multiloop Configurations
Studies on multiloop systems show that one can achieve the lowest flux density noise level when several pickup loops are connected in parallel to form the SQUID inductance. In a DC-SQUID system [1] using 8 pickup loops, a flux density noise level of $1.13\,\text{fT}/\sqrt{\text{Hz}}$ has been detected. Performance capabilities of recently developed, low-temperature DC-SQUIDs are summarized in Table 9.2, and are described in greater detail in Reference [1].

9.3 CAPABILITIES OF VARIOUS SUPERCONDUCTING SQUID MAGNETOMETERS

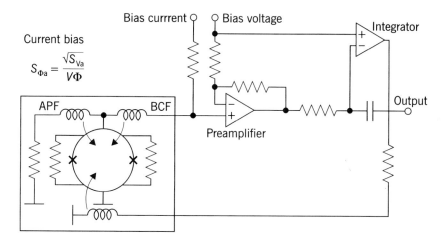

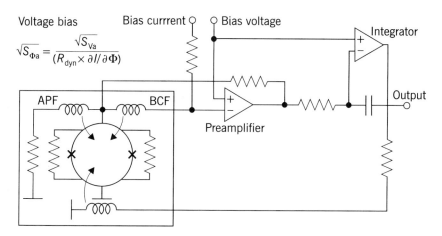

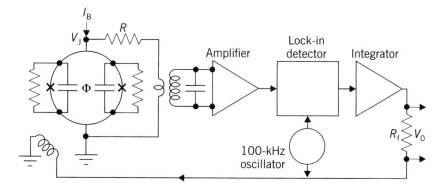

FIGURE 9.10 DC-SQUID sensor using various bias schemes.

TABLE 9.2 Performance Capabilities of Various Low-Temperature DC-SQUIDs [1]

	Type: Bias: Controller:	8-loop Current 2-pole	8-loop Current ?	8-loop Current 2-pole	Washer Current 2-pole	Washer Voltage PI$^{3/2}$	
Chip size		7.2 × 7.2	7.2 × 7.2	4.4 × 4.4	4 × 4	?	mm^2
L		394	$\simeq 600$	206	51	11	pH
C		0.5	$\simeq 0.25$	0.43	1	0.15	pF
L_i		—	—	—	6	$\simeq 1.6$	nH
α		—	—	—	0.85	$\simeq 0.8$	
$16kT\sqrt{LC}$		20	$\simeq 17$	13	10	$\simeq 1.8$	h
ε or ε_c		47	120	108	36	$\simeq 115$	h
$\sqrt{S_\phi}$		2.4	4.7	2.6	0.64	0.5	$\mu\phi_0/\sqrt{Hz}$
$\sqrt{S_B}$		1.13	2.2	4	3.9	—	fT/$\sqrt{Hz}$
f_0		6	25	8	23	15	Hz
$0.14\Phi_0/\sqrt{LC}$		21	$\simeq 24$	31	41	225	μV
$2\delta V$		30	33	47	27	($\simeq 40$)	μV
V_Φ		1.1	0.62	0.66	4.7	($\simeq 0.11$)	mV/ϕ_0
$f_{c,max}$		1	1.2	2.5	0.8	$\simeq 0.6$	MHz
f_c		$\simeq 0.2$	0.04	$\simeq 1.5$	$\simeq 0.15$	0.3	MHz
$\Phi'_{f,max}$		28	67	190	4.8	$\simeq 230$	Φ_0/ms
Φ'_f at $f = f_c/3$		13	?	180	2.7	$\simeq 90$	Φ_0/ms
max. Φ'_f		310	26	5000	19	880	Φ_0/ms
at $f =$		0.85	?	3.1	0.3	1	kHz

9.3.2.3 Washer-Type DC-SQUID Today, a large number of DC-SQUIDs are made in a geometrical configuration developed by Mark Ketchen at IBM, in which the SQUID loop is formed by a large washer (Figure 9.11) that couples the SQUID to the external flux transformer. The washer forms a part of the supercoducting loop containing the JJs and also serves as a one-turn winding on the flux transformer. The dc bias current is applied to one side of the washer and the other side of the SQUID loop is grounded. The washer-type DC-SQUID system shown in Figure 9.12 offers a SQUID loop with low inductance, which provides high magnetic field sensitivity, while allowing efficient coupling to an external flux transformer.

9.3.2.4 DC-SQUID Design for Measurement of Neuromagnetic Signals A highly sensitive DC-SQUID magnetometer with minimum flux noise is required to measure the lowest internal magnetic fields in the brain. A cross-section schematic of a highly sensitive SQUID magnetometer for measuring brain magnetic signals is shown in Figure 9.13. This SQUID system is located inside a small cylindrical, superconducting magnetic shield in the middle of a liquid helium-cooled Dewar. Superconducting pickup coils con-

9.3 CAPABILITIES OF VARIOUS SUPERCONDUCTING SQUID MAGNETOMETERS

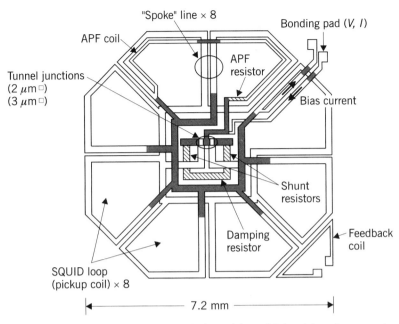

FIGURE 9.11 SQUID magnetometer design with multiple pickup loops to improve sensitivity. APF, additional positive feedback.

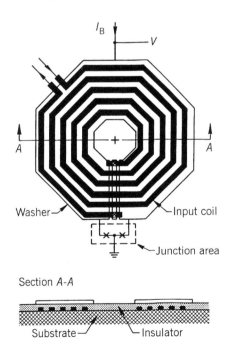

FIGURE 9.12 Schematic diagram of a washer-type DC-SQUID. [Courtesy of Mark Ketchen, IBM Watson Research Center.]

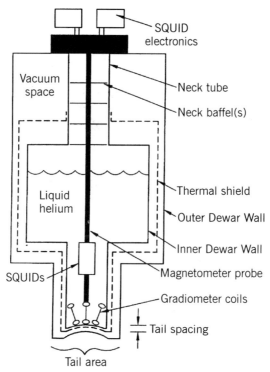

FIGURE 9.13 Cross-sectional schematic of a DC-SQUID magnetometer used to measure brain signals.

figured as gradiometers are maintained at 4.2 K and located at the bottom of the Dewar. The DC-SQUID magnetometer for neuromagnetic diagnosis must be designed for maximum comfort. This system is best suited not only for measurement of brain signals (MEG), but also for heart signals (MCG), stomach signals (magnetogastrogram, MGG), and eye signals (magnetooculogram, MOG). In addition, DC-SQUID devices are widely used in magnetic resonance imaging (MRI) systems, which provide precision measurements of low magnetic fields with significant improvement in S/N ratio compared to RF-SQUID-based systems.

A compact DC-SQUID magnetometer operating at 4.2 K can act like a portable biomagnetic measurement system capable of recording the tangential component of the magnetic field associated with rhythmic alpha activity in the brain. A single-channel SQUID magnetometer system will give a one-dimensional magnetic signal only for a specific location, which may be of little value in the measurement of magnetic fields, particularly in the heart and brain, where several components of the magnetic fields are necessary for reliable clinical assessment and functional evaluation.

9.3.3 Multichannel DC-SQUID Magnetometers

Clinical studies recommend the deployment of the multichannel SQUID magnetometer shown in Figure 9.14 to record several components of the magnetic field simultaneously. This system comprises more than 37 channels, all operating at 4.2 K or lower, and offers highly reliable MCG mapping in a short time (less than 5 min). This kind of system can generate maps that display functional information of the heart without troublesome electrodes, thereby providing a major advantage to cardiologists.

The clinical community and biomedical researchers would like to see a multichannel system with good spatial and temporal resolutions with SQUID noise not exceeding $5\,fT/\sqrt{Hz}$ that can operate ideally in a hospital catheter laboratory. However, cooling requirements for the 4.2-K cryocooler increase in a multichannel system. A multichannel system using SQUID gradiometers can operate in unshielded environments [8] where moderate resolution is acceptable.

The multichannel system [1] shown in Figure 9.14 provides accurate measurements of the three components of the heart vector, the magnetic equivalent of the electronic heart vector. The combination of low-temperature DC-SQUID technology and magnetic shielding room (MSR) can result in a sensitive multichannel MCG detection system with an S/N ratio comparable to that of high-quality ECGs. A multichannel system offers a reliable, safe, and noninvasive technique for diagnosing heart disease.

9.3.3.1 Multichannel Systems Using Second-Order Gradiometer (SOG) for Neuromagnetic Research

The multichannel low-temperature system described in Reference [9] uses 5 DC-SQUIDs to monitor brain activity detected by an array of SQUIDs and 5 RF-SQUID sensors to monitor 3 components of the ambient magnetic field and one component of the gradient required for electronic cancellation of the noise in the 5 signal channels. A SQUID array consisting of N identical SQUID devices coupled in series offers high slew rates. Both the flux noise ($\sqrt{S_\phi}$) and normalized preamplifier flux noise will decrease with $\sqrt{N}$. According to Reference [1], in a sensitive system, 100 identical SQUID devices are required to operate such a series array with a dc flux-locked loop without flux modulation. An array containing 100 SQUIDs has demonstrated signal level measurement capability as low as 2.5 mV, white noise as low as 56 h, and system bandwidth as wide as 7 GHz.

The effectiveness of electronic cancellation of the low-frequency ambient noise in an urban environment was first demonstrated by the neuromagnetic laboratory scientists at New York University in 1982 [8]. The system requires MSR to obtain neuromagnetic data with high accuracy and reliability. This multichannel system operating at 4.2 K demonstrated an intrinsic noise less than $20\,fT/\sqrt{Hz}$ down to a frequency of 0.5 Hz without use of the electronic cancellation method. When operated at 4.2 K or lower it can be important to brain surgeons prior to undertaking complex and risky surgery, because it

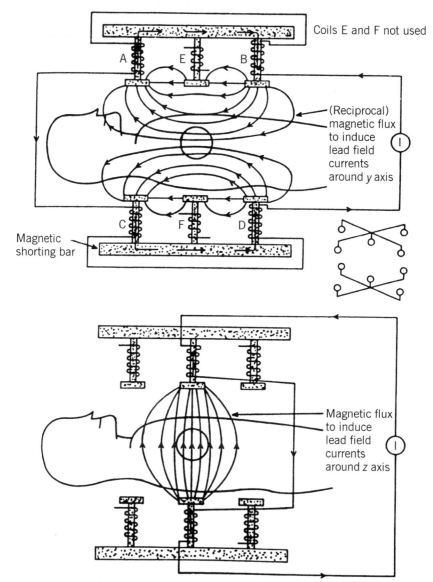

FIGURE 9.14 Multichannel SQUID magnetometer for accurate measurement of three components of the heart magnetic vector [8].

provides 5 simultaneous measurements of the magnetic fields, verification of the signal stability, improved positioning accuracy of the sensor, automatic electronic cancellation of ambient noise components, and risk-free noninvasive MEG measurements.

9.3.4 DC-SQUID Magnetometers for Magnetocardiography

Low-temperature (4.2 K) DC-SQUID magnetometers have potential applications in magnetocardiography. A magnetocardiogram (MCG) mapping reveals functional information of the biomagnetic activity of the heart that is difficult to obtain by other clinical imaging sensors such as magnetic resonance imaging or computer tomography. An electrocardiogram (ECG) measures the electric activity of the heart and can identify any abnormality in the heart. By using DC-SQUIDs, one can generate computer models of propagation excitation in the ventricles. Two- and three-dimensional models using DC-SQUID technology can provide useful information to cardiologists comparing the results obtained through ECGs and MCGs.

In generating ECGs, cardiologists sometimes imagine the cardiac electric forces as the spatial distribution. From body surface ECGs, only a single dipole can be assumed. Furthermore, ECGs detect body surface electric signals coming out through the various layers of different permittivities. On the other hand, MCGs can detect direct vertical magnetic signals from the heart using a low-temperature, second-order gradiometer. This is the most fundamental difference between the ECG and MCG mapping. Finally, MCGs are not as influenced by the organs surrounding the heart.

9.4 SQUID GRADIOMETER SYSTEMS USING LOW-TEMPERATURE SUPERCONDUCTING TECHNOLOGY

SQUID-based gradiometers using low-temperature superconducting technology are capable of measuring the neuromagnetic fields of the brain with high accuracy and reliability. A sensitive (better than $35\,fT/\sqrt{Hz}$), compact SQUID-gradiometer system is most desirable for accurate measurements of the low neuromagnetic fields of the human brain. A first-order gradiometer measures the first derivative of the magnetic flux density along the Z axis, while a second-order gradiometer [5] measures the second derivative of the magnetic flux density along the same axis. Performance capabilities and limitations of some low-temperature gradiometers are briefly described. A gradiometer discriminates strongly against noise sources with small gradients in favor of locally generated signals.

9.4.1 First-Order Gradiometers

Most of the gradiometers made prior to 1980 used both the thin-film and low-temperature (4.2-K) superconducting technologies. A typical gradiometer using both such technologies is shown in Figure 9.15, which can measure off-diagonal or axial gradient of the magnetic flux density. The superconducting flux transformer configurations shown in Figure 9.5 are used in both first-order and second-order gradiometers. The flux transformers are generally made from niobium wire. Using low-temperature and thin-film technologies, one can make an integrated SQUID magnetometer by fabricating a niobium loop across the input coil. Such an integrated DC-SQUID design can provide a magnetic field white noise better than $5\,fT/\sqrt{Hz}$ with a small pickup loop. This low-temperature gradiometer using thin-film technology has potential applica-

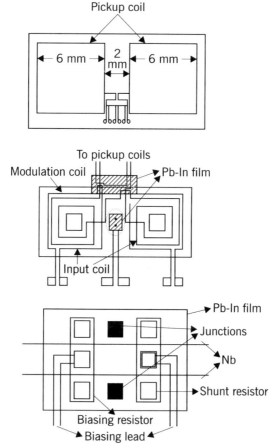

FIGURE 9.15 Critical components used by the low-temperature, first derivative gradiometer [10].

tions in neuromagnetic measurements to detect weak magnetic signals from a human brain.

9.4.2 Magnetically Shielded Second-Order DC-SQUID Gradiometers

A second-order, low-temperature DC-SQUID gradiometer (Figure 9.16) can be used in nuclear magnetic resonance (NMR) probes [9], where high sensitivity and high-quality magnetic shielding are the principal requirements. The magnet, which can be operated in a persistent mode, produces a magnetic field in the direction parallel to the vertical Z axis of the NMR probe. The magnet, the pickup coil, and the sample space are magnetically shielded by an inner superconducting niobium–titanium material and an outer high-permeability metal shielding material. The SQUID is enclosed in a separate superconducting niobium shielding capsule. The second-order gradiometer pickup coil and SQUID detecting circuits are shown in Figure 9.17. This DC-SQUID-based second-order gradiometer demonstrated much higher sensitivity than the RF-SQUID used in the past for NMR application.

9.4.3 RF-SQUID-Based Second-Derivative Gradiometers for Neuromagnetic Field Measurements

RF-SQUID-based second-derivative or second-order gradiometers operating at 4.2 K or lower have accurately measured neuromagnetic signals even in magnetically unshielded environments. The second-order gradiometer shown in Figure 9.18 is based on an improved design of a nondiagonal, first-order SQUID gradiometer. A second-order SQUID gradiometer provides the most effective magnetoencephalographic tool, capable of localizing a source of certain electrical activity in the brain. This sensor is of paramount importance to neurosurgeons, who need to discriminate signals from several spatially separated sources in the cortex. The tangential component of the neuromagnetic field is better for certain types of neuromagnetic sources than the usual measurement of the radial component of the field. Maximum amplitude of the tangential component of the field occurs just above the source, when the detection coil is perpendicular to the current dipole generated by the electrical source. This gradiometer is capable of measuring neuromagnetic magnetic field signals as low as 0.01 nT with high reliability and accuracy. Evoked magnetic fields as low as 0.1 nT have been observed in a transient response. Thus, the RF-SQUID-based second-order gradiometer when maintained at 4.2 K or lower will be most suitable for accurate and reliable measurements of neuromagnetic fields at the head and neck in unshielded environments.

9.4.4 Portable, Low-Temperature SQUID Gradiometers

Portable low-temperature, SQUID-based gradiometers can provide neuromagnetic diagnostic services with minimum cost and complexity. A portable super-

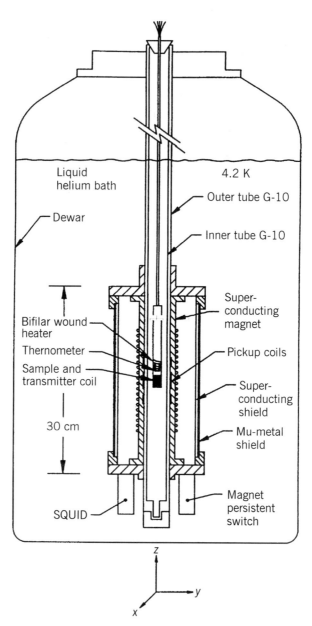

FIGURE 9.16 Magnetically shielded second-order DC-SQUID gradiometer [11].

9.4 SQUID GRADIOMETER SYSTEMS USING LOW-TEMPERATURE ST   237

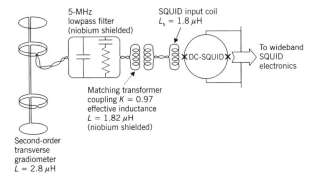

FIGURE 9.17 Schematic diagram of a second-order transverse gradiometer showing pickup coil, detection circuit, and niobium shield.

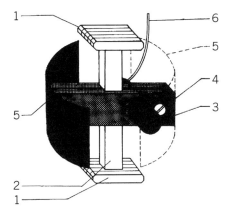

FIGURE 9.18 First-order SQUID gradiometer using low-temperature superconductor technology: 1, quartz slab with flux transformer coils; 2, quartz pillar structure; 3, SQUID; 4, glass filament support; 5, glass fiber springs; 6, coax rf cables.

conducting, nondiagonal SQUID gradiometer [9] offers fast measurement of a neuromagnetic field component parallel to the scalp with sensitivity better than $35 \text{fT}/\sqrt{\text{Hz}}$. A portable SQUID gradiometer design must have a large time interval between the consecutive helium refills, ranging from 16 to 48 h without appreciable increase in weight and size. The portable design must use minimum dimensions for the Dewar assembly to conserve the helium supply, but must be large enough for several gradiometers, which are required for multichannel operation. The small mass of the SQUID gradiometer, due to an improved electronics package and compact Dewar packaging, allows easy manipulation when complex distributions of neuromagnetic fields are measured. Portable gradiometers are best suited for neuromagnetic clinical services in rural areas, where costly magnetic shielding rooms are neither available nor possible.

9.5 SQUID MAGNETOMETERS AND GRADIOMETERS USING HIGH-TEMPERATURE SUPERCONDUCTOR TECHNOLOGY

Magnetometers and gradiometers using low-temperature (4.2-K) superconductor technology are relatively expensive because of the high cost of refrigeration systems. Furthermore, NMR and MRI systems using low-temperature technology suffer [10] from high cost, complexity, and reliability problems. The medical community is demanding open, physician-accessible MRI systems to provide magnetic resonance therapy (MRT) with minimum cost to the patients. However, the physician's demand for unrestricted patient access is incompatible with the MRI system requirements. Nevertheless, the cost of the MRI and MRT systems can be significantly reduced using high-temperature (77 K) superconductor technology. Accelerated development of low-cost refrigeration systems will unquestionably reduce the MRI or MRT system cost to an acceptable level.

9.5.1 DC-SQUIDs

New developments in HTSC devices and circuits make it possible to fabricate DC-SQUIDs using thin films with high critical temperature. Several DC-SQUIDs have been fabricated [11] by laser deposition of YBCO thin films ($T_c = 92$ K) on a strontium–titanate bicrystal substrate. When operated at 77 K these DC-SQUIDs provided acceptable values of critical current, resistance, and voltage modulation produced by an external magnetic field. The white noise energy of the 77-K SQUID with an inductance of 41 pH was around 1.8×10^{-30} J/Hz, which was reduced by two orders of magnitude at 1 Hz using a bias current reversal scheme. Magnetic coupling of a DC-SQUID to a flux transformer demonstrated a white noise of less than $0.036 \, \text{pT}/\sqrt{\text{Hz}}$ at 77 K, which is acceptable for most applications. Low-frequency ($1/f$) noise is an important parameter for certain applications. Low-frequency noise around $10 \, \text{pT}/\sqrt{\text{Hz}}$ at 1 Hz is very high, but can be reduced with HTSC thin films of high quality.

The 77-K operation of the SQUID cannot compete in sensitivity with that at 4.2 K, but its low system cost and simplicity are very appealing. HTSC technology can be used in the design of planar gradiometer to give sufficient sensitivity for certain applications, such as nondestructive evaluation. If the sensitivity of a SQUID-based gradiometer using HTSC technology is high at frequencies down to 0.1 Hz, it will be most attractive for neuromagnetic applications, where cost and simplicity are the principal requirements. High-temperature DC-SQUID magnetometers and gradiometers [12] have been fabricated using thin-films of YBaCuO (YBCO) and PrBaCuO (PBCO) on MgO substrates. These magnetometers demonstrated a sensitivity as low as $68 \, \text{fT}/\sqrt{\text{Hz}}$ at 77-K operation. A first-order gradiometer using HTSC technology provided a sensitivity at 1 Hz better than $95 \, \text{fT}/\sqrt{\text{Hz}}$ at 77 K. These sensitivity levels can be improved if the input cells of the flux transformers are mounted in a flip-chip configuration.

A SQUID-based gradiometer using two HTSC octagonal washers offers significant advantages over trilayer devices in terms of sensitivity, lower production cost due to high yield, robustness in trilayer structures, and self-shielding from external fields. It is expected that HTSC SQUID magnetometers and gradiometers will offer NMR, MRI, and MRT systems with unrestricted access to patients and physicians, in addition to lower cost and less complexity at 77 K operation. By mapping neural activity with HTSC SQUID magnetometers, one can locate electrical regions in the brain within a few millimeters with minimum cost. Furthermore, multichannel HTSC SQUID systems will allow neuromagnetic field measurements in the brain at more than 37 different sites simultaneously.

SUMMARY

Integration of low-temperature and high-temperature superconductor technologies to clinical diagnostic equipment will improve sensitivity, resolution, and patient comfort. Cryogenically cooled medical equipment offers a substantial reduction in noise, weight, size, and power consumption. SQUID-based magnetometers and gradiometers when operated at 4.2 K have demonstrated significant improvement in the overall performance of medical equipment such as MRI and CT scanners. MCGs, MEGs, and MGGs provide improved and reliable alternative tools for detecting diseases associated with brain, heart, and stomach, respectively. Cryogenically cooled, multichannel, second-order gradiometers are widely used to measure heart and brain activities. Magnetocardiograms seem to offer more reliable and accurate measurement of heart activities than electrocardiograms.

REFERENCES

1. D. Drung. Recent low-temperature SQUID developments. *IEEE Trans. Appl. Superconductivity* 4:121–126 (1994).
2. K. Harakawa et al. Evaluation of high-performance magnetically shielded room for biomagnetic measurements. *IEEE Trans. Magn.* 32(6) (1996).
3. J. P. Wikswo, Jr. SQUID magnetometer for biomagnetism and nondestructive testing: important questions and initial answers. *IEEE Trans. Appl. Superconductivity* 5:74–92 (1995).
4. K. Hara. *Superconducting Electronics.* Englewood Cliffs, NJ: Prentice Hall, 1987, p. 157.
5. J. Clarke. Principles and applications of SQUIDS. *Proc. IEEE* 77:1208–1221 (1989).

6. K. Shirae et al. Vector SQUID magnetometer for biomagnetic measurements. In K. Hara. *Superconducting Electronics*. Englewood Cliffs, NJ: Prentice Hall, 1987, pp. 172–173.
7. N. Fujimaki et al. 8-channel array single-chip SQUID connection to a Josephson multiplexer. *IEEE Trans. Appl. Superconductivity* 23:2601–2604 (1993).
8. E. Weinberg, S. J. Williamson et al. Five-channel SQUID installation for unshielded neuromagnetic measurements. In *Biomagnetism: Applications and Theory*. Elmsford, NY: Pergamon, 1995, pp. 46–51.
9. J. Anderson et al. DC-SQUIDs for NMR probe. *IEEE Trans. Magn.* 32:262 (1996).
10. C. Weggel et al. Redesigning open, interventionally MRI magnetometer. *Superconductor Industry*, 16 (Fall 1992).
11. A. H. Miklich et al. Bicrystal YBCO DC-SQUID with low noise. *Applied Superconductivity Conference*, 1992, p. 65.
12. M. N. Keene et al. Low-noise HTS gradiometers and magnetometers constructed from YBaCuO and PrBaCuO thin films. *IEEE Trans. Appl. Superconductivity* 5:2923–2926 (1995).

CHAPTER TEN

Application of Superconducting Technology to Generators, Motors, and Transmission Lines

The application of superconducting technology to generators, motors, propulsion systems, and transmission lines is discussed. Studies performed in early 1970s indicate that implementation of superconductor technology in these systems offers significant reduction in weight, size, and cost and considerable improvement in efficiency. Maximum benefits of this technology can be realized in ship-propulsion systems, electrodynamic levitation systems, and high-power turbogenerators. These benefits are due to two attributes:

1. Superconductors carry huge currents with no dissipation and eliminate iron in some parts of the machine, thereby reducing weight and size.
2. Superconductors can generate large magnetic flux densities, which will lead to considerable reduction in the physical size of the electrical equipment.

In metallic superconductors such as niobium–titanium alloy (NbTi) deep cryogenic cooling at 4.2 K is required, which is expensive. High-temperature superconductors using rare-earth or ceramic compounds provide cooling at 77 K or lower with minimum cost, but with improvement of 2 orders of magnitude in current-carrying capability instead of 20 orders of magnitude at 4.2 K. In brief, the use of superconductor technology makes electric power equipment smaller, lighter, and more efficient, in addition to improving dynamic response. High-temperature superconductors with sufficient mechanical strength and large current-carrying capability will have a significant impact on the performance of motors, turbogenerators, and transmission lines.

Based on the studies performed by C. J. Mole et al. [1], it was concluded that liquid-metal current collection systems are needed for very high-power machines. The ability of superconductors to generate a strong magnetic field with minimum power consumption in a compact envelop has been demonstrated. Conventional windings, if used, would require large input power, thereby adversely affecting the weight, size, and cost of the equipment.

10.1 DEVELOPMENT HISTORY OF SUPERCONDUCTING MACHINES

Although the phenomenon of superconductivity has been known for more than 65 years, its practical application to electrical machines has happened only during the last 30 years. The development of the intrinsically stable superconductor for a large pulsed magnet was a first major step leading to the integration of superconductor technology in high-power electrical and mechanical machines. The discovery of niobium–titanium alloy provided the first practical superconductor for high–power machines. State-of-the-art multifilamentary superconductor elements are now widely used in current superconducting machines.

10.2 ADVANTAGES OF SUPERCONDUCTING MACHINES

A conventional synchronous machine (Figure 10.1) employs two basic configurations: a round rotor machine and a salient pole machine. The round rotor design is used for high-speed power generators, while the salient pole configuration is used for slow power generators. The interaction between the magnetic fluxes generated by the armature and rotor windings develops an electromagnetic (EM) torque. This torque opposes the rotation of the shaft and, thus, requires a mechanical driving torque from the prime mover such as steam turbine or a gas turbine or oil engine to sustain rotation. The EM torque is the practical mechanism through which greater electrical power output is feasible with greater mechanical power input.

In a conventional motor, ac power is applied to the armature winding and dc excitation is applied to the field winding. The magnetic fields produced by the stator and rotor windings must be constant in amplitude and stationary with respect to each other to produce a steady EM torque. The torque in a motor is in the direction of rotation and balances the opposing torque needed to drive the mechanical load. The rating of a conventional machine is limited by certain requirements imposed by the operating parameters, such as speed, voltage, reactance, and cooling. The rating of a superconducting machine is limited by the complexity of the mechanical structure and the operating temperature limitiations imposed by the insulation materials.

In large turboalternators, the ratings have been significantly improved by using advanced cooling schemes, but at the expense of higher losses in the field windings due to increased current densities. These losses are consider-

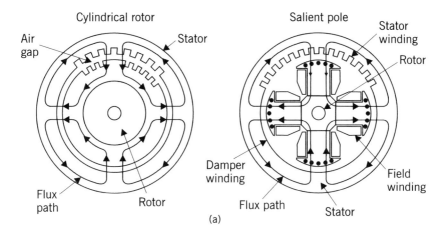
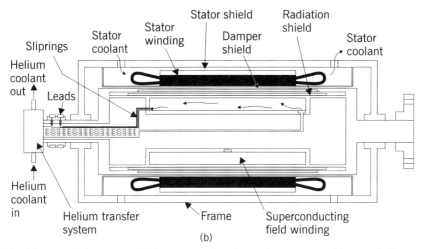

FIGURE 10.1 Elements of (a) a conventional synchronous machine and (b) an ac superconducting generator [1].

ably reduced in case of superconducting turboalternators. Cryogenic operation allows an increase in the flux density linking the stator windings by a factor ranging from 3 to 4, in addition to increased space which offers a further hike in rating by a factor of 6 to 10 at the stator windings. Cryogenic operation offers significant improvements in size, weight, power system stability, efficiency, and capital plant cost, particularly in large machines. The current-carrying capacity in superconductors is dependent on both the operating temperature and magnetic flux density: the higher either of these quantities, the lower the current density. Significant reduction in weight and size of a superconducting machine [1] is evident from data shown in Table 10.1.

TABLE 10.1 Relative Weight for Conventional and Superconducting Machines

Types of Machine (rating)	Conventional	Superconducting
Turboalternator (1000 MVA)	1	0.10
Turboalternator (100 MVA)	1	0.37
Ship-propulsion generator (25 MVA)	1	0.12
Ship-propulsion motor (3500 hp)	1	0.15

10.2.1 Current Capability and Rating

The output rating of a superconducting machine is strictly a function of the current-carrying capability of the superconductor used and its operating temperature (Figure 10.2). The current density shown in Figure 10.2 is for the best water-cooled copper winding, whereas the current density and magnetic flux density data in Figure 10.2 are for superconducting field windings. The reduction in size of a superconducting machine is defined by an output reflection coefficient, which can be written as follows:

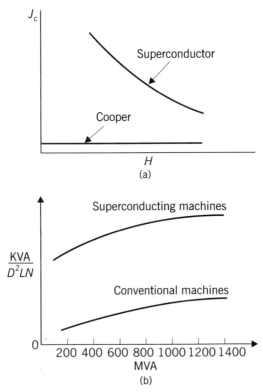

FIGURE 10.2 (a) Current densities for copper and a superconductor. (b) Output coefficient for conventional and superconducting machines [1].

$$R_{out} = \frac{KVA}{D^2LN} \tag{10.1}$$

where KVA = power rating of the superconducting machine
D = stator diameter
L = core length
N = synchronous speed of the machine

The reflection coefficient (Figure 10.2) is a true indicator of the power output capability of the machine. The magnitude of this parameter increases with cryogenic operations, which offers higher output power with reduced size.

10.2.2 Improvement in Efficiency at Cryogenic Temperatures

Significant improvement in efficiency is the potential advantage of a superconducting machine, because of the absence of I^2R losses at cryogenic temperatures. In many cases iron is no longer required, which will permit an extra amount of copper windings in the space formally occupied by the iron. This will increase the magnetic flux density, thereby further improving both the power output capability and efficiency of the machine.

10.3 SUPERCONDUCTING SYNCHRONOUS GENERATOR

This superconducting machine has three windings: an armature winding, a dc field winding, and a damper winding interposed between the armature and field windings. The requirement of the superconducting field winding to be maintained at 4.2 K is satisfied by four basic synchronous machine elements. Axial and lateral cross-sectional views of a superconducting generator are shown in Figure 10.3. Specific details on field winding, damper winding, and thermal radiation shield for superconducting generator are shown in Figure 10.4. The synchronous machine torque is proportional to the armature and field currents. At cryogenic temperatures, the field current is much higher, leading to a significant increase in flux density and flux linkages. Because of the extremely high torque in a superconducting machine, the rotor assembly must be designed to meet the stringent thermal and structural requirements at 4.2 K.

The field winding in a rotor must be continuously supplied with liquid helium to maintain the required operating temperature under all load conditions. The helium, after passing through the field windings, is used to cool the electrical leads and slip rings and is then returned to the refrigeration system for recycling with minimum loss of the refrigerant. Specific details on cooling schemes for field windings are shown in Figure 10.5.

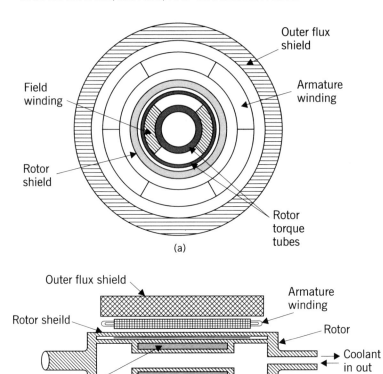

FIGURE 10.3 Cross section of a superconducting generator: (a) axial view; (b) side view.

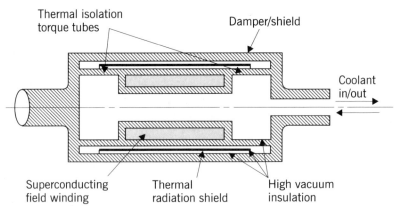

FIGURE 10.4 Superconducting field winding and associated thermal and damper shields [1].

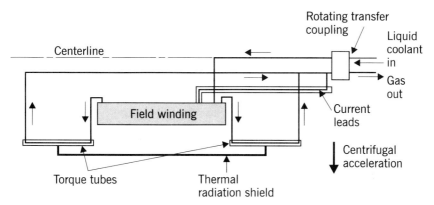

FIGURE 10.5 Cryogenic cooling scheme for the superconducting field winding and current leads.

10.3.1 Thermal Radiation and Damper Shields

Axial radiation shields (Figure 10.5) are required to reduce the direct thermal radiation from the ambient structures to specified low operating temperature required to cool the magnet support structure and other cold components. The thermal radiation shields are cooled by the exhaust helium at a specified temperature over the 20- to 100-K range.

Cylindrical damper shields around the field windings are necessary to provide restoring torque to the rotor during the unexpected rotor swings and to shield the field windings from the ac fields generated by the stator winding harmonics and negative sequence fields generated by the steady-state and transient-state conditions. Due to high electrical conductivity of copper and aluminium conductors at cryogenic temperatures, very thin shields with small losses can be used to achieve effective shielding. Under transient conditions the losses in the damper shield can represent a large load for the helium cooling system, which must be considered during the selection of a cooling scheme. Optimum locations of EM damper shield, thermal radiation shield [6], and magnetic shield for a superconducting generator are shown in Figure 10.6.

The damper shield must be designed to withstand torques under sudden short-circuit conditions. The use of a damper shield will put short-circuit torques on the ambient structures, thereby limiting the torque on the cryogenically cooled field winding. The outer rotating Dewar shell remains at the ambient temperature, but the inner cold field winding under cryogenic temperatures will shrink in length relative to the Dewar shell. Special provisions must be made for this shrinkage to maintain low heat leakage as well as rotor dynamic stability [1] under various electrical load conditions.

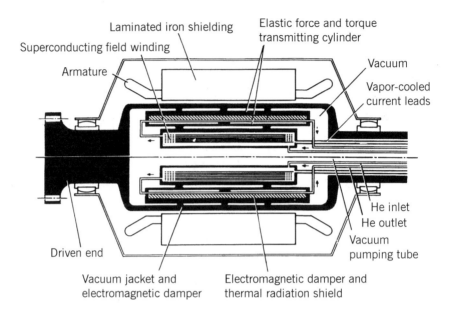
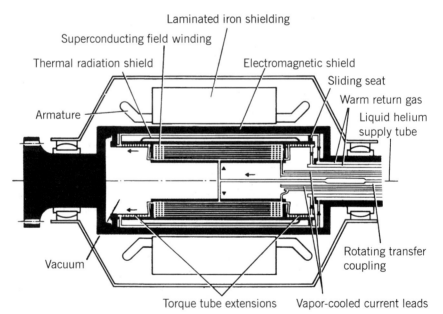

FIGURE 10.6 Schematic views of a superconducting generator showing the locations of various shields [6].

10.3.2 Stator Windings

The stator of a superconducting synchronous generator comprises a stator winding and a shielding structure to provide magnetic shielding, mechanical support, and cooling provisions. The stator windings will be subjected to ac fields, which will introduce eddy-current losses depending on the operating temperature and peak surface flux density. According to Reference [2], the ac losses ($\mu W/cm^2$) increase with the increase in operating temperature and peak surface flux density. Typical ac losses in a Nb_3Sn superconductor as a function of temperature and peak surface flux density are shown in Figure 10.7. Implementation of an eddy-current shield will not only improve the magnetic coupling between the magnetic field and stator winding, but also will attenuate the rotating field. The eddy-current shield for a superconductor generator must be constructed with a material of high electrical conductivity, such as copper.

10.4 DESIGN ASPECTS FOR EFFICIENT SUPERCONDUCTING MACHINES

The principal objective of using superconductor technology is to achieve the highest efficiency of the machine, which will offer maximum savings in weight, size, and operating costs. During the 1960s and 1970s experimental investigation was carried out on small superconducting ac machines. These machines utilized dc superconducting field winding in the stator and armature winding in

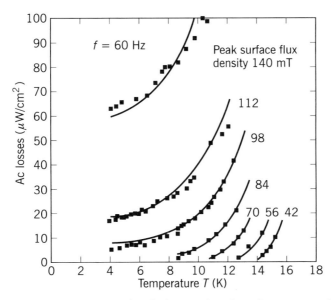

FIGURE 10.7 Ac losses in Nb_3Sn as a function of temperature [2].

a conventional rotor. A major step forward was achieved with a design that demonstrated the feasibility of rotating superconducting field winding [3] for the superconducting machine (Figure 10.8). This design not only reduced the weight, size, and complexity of the rotor, but also significantly improved the efficiency of the machine. It allows the machine reactances to be charged by varying the armature winding thickness, thereby yielding economical design of the armature and optimum location of the damper shield. An optimum design of a machine with effective eddy-current shield requires a large shield diameter to achieve high efficiency, which is practical only when the armature windings are in the stator.

10.5 CRITICAL ELEMENTS OF LARGE SUPERCONDUCTING AC MACHINES

The Westinghouse Superconducting Electrical Machine Group designed and tested several large machines in 1970s with ratings exceeding 5 MVA. These

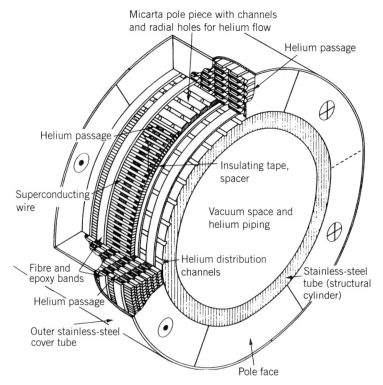

FIGURE 10.8 Helium flow in various sections of high-power dc superconducting field winding.

10.5 CRITICAL ELEMENTS OF LARGE SUPERCONDUCTING AC MACHINES

superconducting machines with field windings on the rotor were designed for central power-generating stations for interrupt-free operations and with maximum load factors. During the development of superconducting synchronous ac machines to the primary focus was on significant improvement in the power-to-size ratio and efficiency, which are recognized as the principal requirements of large turboalternators for central power stations. The use of superconducting field windings was later extended to ship-propulsion systems and aircraft generators.

Large turboalternators require low machine reactances for good transient stability [6]. The maximum possible output for a given size of the machine is determined by the electrical and mechanical limits. The electrical limits are relatively easy to satisfy compared to mechanical limits. The mechanical limits are established by centrifugal forces, bending stresses, torsional vibrations, and bending vibrations at the critical speeds of the turboalternators. Material strength and operational stability [5] limit the maximum rotor diameter as well as the rotor length-to-diameter ratio. These limiting factors must be taken into account in the design of large superconducting ac machines. If the demand for electrical power consumption continues to grow, the developing countries, including the United States, Germany, Switzerland, the United Kingdom, Russia, and Japan, will be compelled to deploy large superconducting ac machines for large central power stations.

10.5.1 Potential Advantages

Superconducting generators used in central power stations offer potential advantages such as significant reduction in weight and size, higher efficiency, improved machine rating, reduction in plant cost, and remarkable improvement in power stability.

10.5.2 Critical Design Aspects

The rotor of a superconducting generator contains the suspended superconducting field windings. The suspension must transmit large torques from the warm rotor structure to the cold superconducting field winding to minimize the heat intake from ambient temperature to the low-temperature region. The torque to the superconducting field winding is transmitted from the rotor shaft at ambient temperature by several radial supports which are thermally isolated via a stainless steel cylinder (Figure 10.8). The heat intake by conduction must be further reduced through installation of a thermal shield ideally maintained at a temperature range of 50 to 100 K. Specific details on the critical components and various shields for a typical ac superconducting generator are illustrated in Figure 10.9. The EM damper shields must be designed to reduce ac magnetic field intensity and current amplitude experienced by the field winding due to phase unbalance, stator winding harmonics, and transient

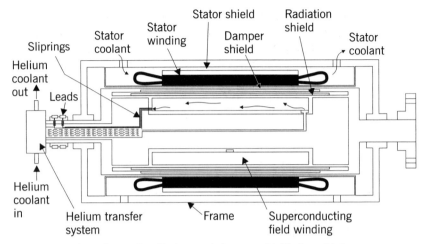

FIGURE 10.9 Locations of radiation and damper shields in a high-power ac superconducting generator.

operation. The time constant of the EM shield winding needs to be long enough to provide effective shielding for superconducting field winding.

10.5.3 Cooling Requirements

Cooling requirements present serious problems, particularly for large ac superconducting generators. Heat conduction by the suspension system, Joule heating, thermal conduction in the current leads, and ac losses contribute to the overall heat load of the superconducting field winding. Losses in the suspension structure and EM shields must be kept to a minimum to retain high efficiency as well as to minimize helium consumption. For example, in a 3000-MVA superconducting generator equipped with warm and cold electromagnetic shields the losses under steady-state operation are typically below 20 kW at the cold shield (80 K) and about 100 W at the field winding, including the losses in the current leads. These losses correspond to a refrigerator input power of about 600 kW, which amounts to only 0.2% ($0.02 \times 3000/100 = 600$ kW) of the generator power.

Transient heat generated during the short circuit contributes to transient heat load due to rapid field variations. However, in the case of a rare short-circuit condition, it is possible to absorb the heat energy released by the heat capacity of helium cooling the field winding without a rise in the critical temperature. This means that additional heat generated during the short circuit must be taken into account during the design of a cooling system for the EM shields. Typical electrical losses in the rotor of a central power station superconducting generator are summarized in Table 10.2. Maximum losses occur under unpredicted sudden short-circuit conditions.

TABLE 10.2 Losses in the Rotor of a Superconducting 1000-MW Generator at 4.2 K Operation [6]

Mode of Operation	Data	Duration t of Time Constant τ (s)	Maximum Loss in the Conductor (kW/m³)	Electrical Losses				Total Dissip. Energy (kW)	Evaporated Amount of He (L)
				Mean Losses in the Winding (W)	Mean Losses in the Slotted Support Cyl. (W)	Mean Losses in the Cold Damper (W)	Mean Losses in the Support Cyl. (W)		
Charging		$t = 337$	2.3	27	?	50	?	25 (?)	8.6 (?)
Rapid discharge after load shedding		$t = 7$	5.7	461	?	10,000	?	70 (?)	24 (?)
Sudden short circuit	$B_w = 10\,\text{mT}$	$\tau = 0.8$	1.5	480	58,000	172,000	6400	95	33
Hunting	$\theta = 70°$ $f = 1\,\text{Hz}$	$\tau = 5$	3.3	1000	1000	121,000	6000	325	110

Note. Electrical losses in the main components of the rotor of a 1000-MW superconducting generator at transient operation.
Unit Symbols: s = second; W = watt; kW = kilowatt; L = liters; cyl = cylinder.

10.5.4 Performance and Cost Comparison for Large Conventional and Superconducting Generators

Performance parameters and projected cost of ownership for 2-pole and 4-pole large conventional and superconducting turboalternators [6] are summarized in Figure 10.10. The economic crossover between the large, conventional and

	1200 MVA[a] 2-pole conventional	1200 MVA[a] 2-pole superconducting	1500 MVA[b] 2-pole conventional	3000 MVA[b] 2-pole superconducting
Rated voltage (kV)	26	34	27	27
Armature outer diameter (m)	4.3	3.7	4.3	4.3
Total length (m)	13	7.2	15	11
Total mass (kg)	630	140	580	480
Synchronous reactance x_d (%)	181	52	220	90
Transient reactance x'_d (%)	32	29	31	70
Subtransient reactance x''_d (%)	26	16	23	55
Efficiency (rated load) (%)	98.6	99.4	98.75	>99

[a] Data from Westinghouse.
[b] Data from KWU.

(a)

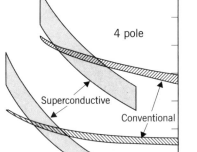

(b)

FIGURE 10.10 (a) Comparison of superconducting and conventional turbogenerators. (b) Comparison of evaluated cost for conventional and superconducting steam turbine-generator sets [7].

superconducting generators seems to be around 1300 MVA for 2-pole generators and roughly 1700 MVA for 4-pole generators. Optimum design of a large superconducting generator has a breakeven point of around 500 MVA. These data indicate significant savings in capital costs and improvement in performance level for large superconducting generators. The efficiency enhancement is about 0.6 to 0.8% for a 2-pole superconducting turbogenerator and 0.3 to 0.5% for a 4-pole superconducting generator.

10.5.5 History of Superconducting Turbogenerator Projects

The Electric Power Research Institute in the United States predicted in 1970 a total U.S. market for superconducting generators of about $350 millions by 1975, increasing to more than $2 billion in the year 2000 based on 1970 dollars. Preliminary studies performed in Europe indicated that about 200 superconducting generators, each with a rating of 3000 MVA, will be needed in Europe by the year 2000. The studies further indicated that demand for 1000-MVA generators may run into several hundreds.

Based on these predictions, industrial nations, such as the United States, England, France, Germany, Japan, Switzerland, and Russia, engaged in the research and conceptual design studies on superconducting generators. Chronological history of state-of-the-art prototype superconducting generators developed by various nations is shown in Table 10.3. At the beginning of the research studies and experimental investigations, primary focus was on the development of individual components—current leads, cooling system elements, transfer coupling mechanism, superconductor filament matrix, and winding models.

Westinghouse and General Electric in the United States carried out initial conceptual design studies on 2-pole superconducting generators with ratings from 300 to 1200 MVA. Western European countries performed design studies in late 1970s on machines with ratings varying from 500 to 3000 MVA. A prototype superconducting generator of 300-MW capacity was tested by Westinghouse in 1984, a prototype 1350-MW generator was tested by the former USSR in late 1995, a prototype 1000-MW generator was tested by France in 1988, and a prototype superconducting generator with 1000-MW rating had a test plan by West Germany in late 1988. Fuji Electric and Mitsubishi Electric of Japan have been working on superconducting machines since 1982, but reliable data on the development status of large generators are not readily available.

10.6 SUPERCONDUCTING GENERATORS FOR AIRBORNE APPLICATIONS

Preliminary studies performed by Westinghouse indicated that the airborne superconducting generator design must deploy rotating superconducting

TABLE 10.3 State-of-the-Art Superconducting Generator Development Programs Undertaken by Various Countries

Country	Institution	Conceptual Design Studies	Model Generators Components		Prototype Generator	
			Power (MW)	Year of Test	Power (MW)	Year of Test
USA	MIT	2000	0.045	1969	?	?
			3	1972		
			10	1980/81		
	Westinghouse	300/1200	5	1972	300	1984
	General Electric Co.	300/1200	20	Rotor:1980 Compl.: 1981	?	?
Russia	U.S. Dept. of Energy	300/1000	>150	1992	100	1993–1996
	University Institute of Electrical Machinery Leningrad	2000	1	1972	350	1987
			20	1980?	1350	>1990 (planned)
France	CGE/Alsthom|EdF	1000/2000	Rotor modified	1980	500–1000	Before 1990
Japan	Mitsubishi/Fuji	—	6	1977	—	—
	Toshiba	200	30	1982	—	—
	Hitachi	—	50	rebuilt		
	Super-GM Program	250	70	1995	70	1996–1998
China	Shanghai PERI and EMMW	?	0.4	1974	?	?
		?	0.4	1977	?	?
England	IRD/CERL/GEC	500/1300	Austenitic steel forging	1979	120	
Switzerland	BBC	1000/3500	?	?	?	?
Italy	Ansaldo	?	Rotor modified	?	?	?
Germany	University of Munich		0.3	1980		
	KWU/Siemens	1000/2000/3000	Rotor modified	1982	500–1000	1988 (planned)

Source: Proceed. IEEE, 77(8), August 1989; IEEE Spectrum, July 1997.

NbTi-multifilament field windings embedded in the rotor assembly. Prototype generators with ratings of 1 to 20 MVA were successfully operated in the United States in early 1980. Westinghouse built a 5-MVA superconducting generator for airborne application, which provided reliable data under normal operations and short-circuit conditions. The All Union Research Institute for Electrical Industry of the former USSR tested a 1-MVA generator in 1975 with a projected capability of 20 MVA by 1985. Westinghouse was deeply involved in the design and development of several superconducting machines over the 1970–1973 period, which included a 5-MVA, 4160-V, 3600-rpm, 60-Hz generator and a 5-MVA, 500-V, 1200-rpm, 400-Hz generator. Westinghouse also investigated the use of a Nb_3Sn superconductor and concluded that this superconductor offers lower cost, smaller size, higher critical temperature, and higher operating temperature (4.2 K or slightly higher).

The airborne superconducting machine [4] must be maintenance-free, small, lightweight, and reliable. Conventional oil-cooled aircraft generators have a potential specific mass of 0.6 lb/KVA with a 5-KVA rating, which can be reduced to 0.4 lb/KVA with a 20-MVA rating using the superconducting field windings. These claims can be verified from the curves shown in Figure 10.11.

10.7 SUPERCONDUCTING MOTORS

Studies performed in early 1970s indicated that the efficiency benefit of superconducting motors are not significant. It is the size advantage of the superconducting motor that makes it attractive for industrial application. Compact superconducting motors have potential applications in electrical propulsion systems for high-performance ships, such as naval high-speed combatants, and high-speed rail vehicles.

Two classes of superconducting motors are of great interest: ac synchronous motors and dc homopolar motors. Operational capabilities and limits of conventional and superconducting motors are summarized in Figure 10.12. Ac synchronous motors are similar to ac synchronous generators in form, but require electrical power to provide mechanical power at the shaft. In a dc homopolar motor, also known as an acyclic (Figure 10.13), the superconducting magnet is a stationary solenoid and its flux passes radially through a series of rotating and stationary current-carrying drums. The conductors are continuous cylinders carrying currents in the axial direction. The interaction between the axial current and radial magnetic field generates an azimuthal force called the output torque. One of the best features of the dc homopolar machine is the complete absence of a demagnetizing effect.

Most of the developmental efforts on dc homopolar motors have focused on slip ring and collector technology. The dc homopolar machine, which has demonstrated the longest operation, was built by the United Kingdom in 1972. This particular machine was developed based on Faraday disk design

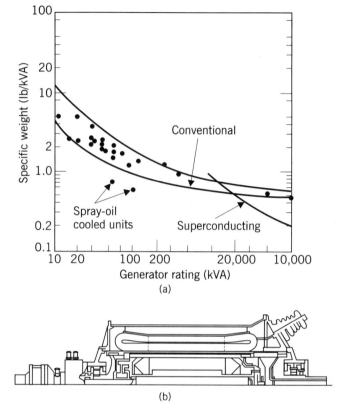

FIGURE 10.11 (a) Specific weight and (b) cross-sectional view of a superconducting airborne generator.

(Figure 10.14). A superconducting homopolar machine using a drum configuration is shown in Figure 10.15. Specific details on these homopolar machines are available in Reference [3].

A typical dc hompolar 3250-hp, 200-rpm motor demonstrated an efficiency of better than 90% at 4.2 K, which is practically impossible with a conventional motor. The dc homopolar motors are best suited for rail transportation and ship propulsion. Conventional motors are very heavy and are practical for ship propulsion only with costly and heavy reduction gear mechanisms.

10.8 APPLICATION OF SUPERCONDUCTING SYNCHRONOUS MACHINES FOR SHIP-PROPULSION SYSTEMS

Studies performed during the 1970s reveal that a dual-armature superconducting machine consisting of a stationary and rotating armature and a torqueless

10.8 SUPERCONDUCTING SYNCHRONOUS MACHINES FOR SHIP-PROPULSION

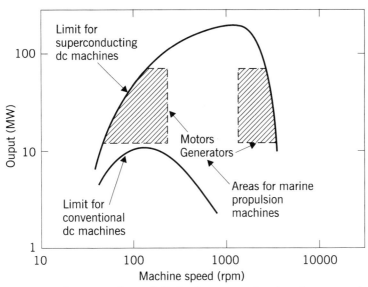

FIGURE 10.12 Output and speed limits for conventional and superconducting dc machines and motors [1].

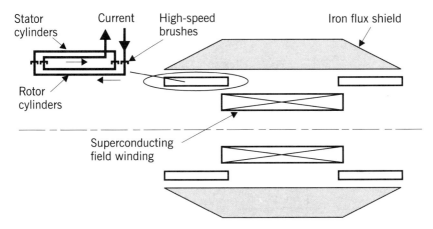

FIGURE 10.13 Specific details on superconducting field winding and iron flux shield for a high-power dc homopolar superconducting motor [3].

rotating superconducting field winding is best suited for ship-propulsion systems. A ship-propulsion system requires low-speed, multipole, superconducting synchronous machines for long-duration optimum performance. The ac ship drive mechanism is most attractive for applications where controllable pitch propellers are mandatory, such as high-speed surface effect ships or hydrofoils, and where constant cruising speed is the principal requirement, as in the case of a merchant ship. A synchronous electrical machine is highly

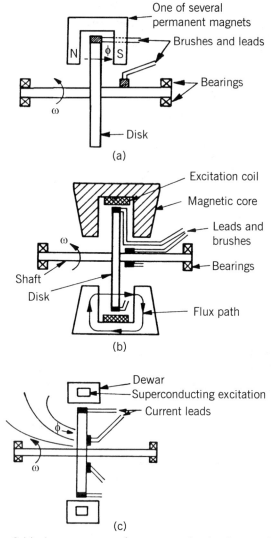

FIGURE 10.14 Critical components of a superconducting homopolar machine based on a Faraday disk [3]: (a) Faraday disk; (b) basic homopolar machine; (c) superconducting homopolar disk machine.

favored by ship operators because of very long life and reliable operation over long distances. Both dc and ac superconducting motors [9] are suited for ship-propulsion systems. Dc superconducting motors are widely used by small, highly maneuverable ships because of large weight and size. Typical performance and physical parameters for superconducting ac and dc motors are summarized in Table 10.4.

10.8 SUPERCONDUCTING SYNCHRONOUS MACHINES FOR SHIP-PROPULSION

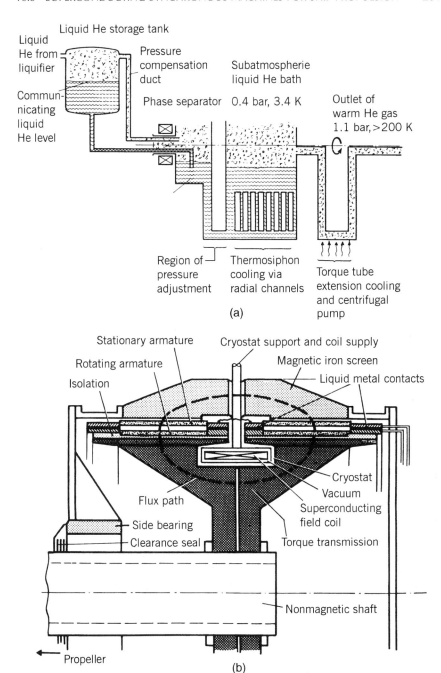

FIGURE 10.15 Cooling scheme for superconducting field winding of a homopolar motor with drum configuration: (a) principle of the cooling system of a rotor with superconducting field winding; (b) schematic of a superconducting homopolar motor of the drum type.

TABLE 10.4 Superconducting 30-MW Dc and Ac Motors for Ship Propulsion

Parameters	Superconducting Motor	
	Dc	Ac
Speed (rpm)	180	200
Shaft torque (KN-m/lb-ft)	$1600/1.2 \times 10^6$	$1430/1.1 \times 10^6$
Operating voltage (V)	300 (dc)	6900 (3-phase)
Current requirement (kA)	100	2.5
Efficiency (%)	98	97
Diameter/length (m)	1.86/3.70	1.73/2.35
Weight (kg/lb)	37,500/82,500	15,342/33,752

10.9 SUPERCONDUCTING ELECTRODYNAMIC LEVITATION SYSTEMS

Magnetic levitation and propulsion are the basic requirements for tracked high-speed ground transportation systems with speeds exceeding 500 km/h over distances from 200 to 2000 km. Low noise level, no environmental impact, high speed, and improved reliability are the major advantages of a superconducting levitation system. Preliminary operating cost estimates for levitated trains indicate equal or better economics when compared to air travel costs.

10.9.1 Operating Principle of a Levitated Train

The operation of an electrodynamic levitated train requires the superconducting magnets to be arranged longitudinally on both sides in the lower part of the vehicle (Figure 10.16). The concrete track is equipped with continuous aluminium sheets laid out in the longitudinal direction just below the superconducting magnet. Loops or coils with appropriate dimensions can be used instead of aluminum sheets. The flux generated by the superconducting magnet penetrates the aluminum sheets vertically down and induces currents in the aluminum sheets, when the vehicle is moving in the longitudinal direction. The interaction of these currents with the currents in the superconducting coils generates a repulsive force, which lifts the vehicle above the track. At very low speeds (less than 50 km/h) or at standstill, the lifting force is practically zero and the vehicle has to be supported by the wheels on the track. The induced currents in the aluminum sheets also generate small drag forces which need to be overcome by the propulsion system.

10.9.2 Superconducting Propulsion System

An air-cooled linear synchronous motor with superconducting field windings is best suited for the levitation propulsion system. The superconducting field windings are arranged longitudinally in the middle of the lower part of the

10.9 SUPERCONDUCTING ELECTRODYNAMIC LEVITATION SYSTEMS

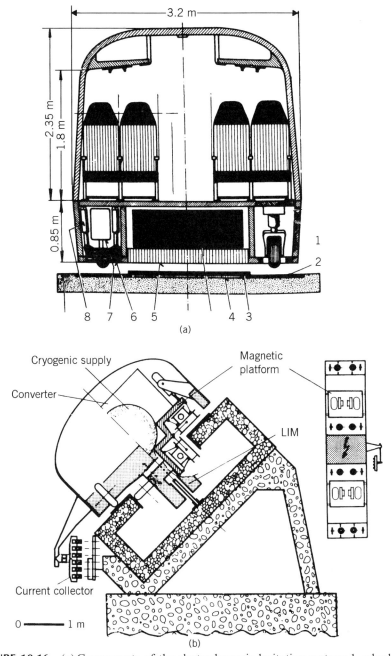

FIGURE 10.16 (a) Components of the electrodynamic levitation system: 1, wheel for low-speed suspension; 2, aluminium sheet for levitation; 3, guidance coils; 4, LSM coils; 5, propulsion superconducting magnet; 6, levitation superconducting magnet; 7, liquid helium storage container; 8, secondary suspension. (b) Components of the levitation system with a 45° inclined test track.

vehicle. A multiphase armature winding with a meander-shaped cable is horizontally laid into the track and is fed sectionally (typical section length is less than 5 km) with a current of variable frequency by the converter substations. The armature winding generates a traveling EM wave with variable speed along the track section supplied by the variable frequency current. The interaction between the superconducting field coils and the traveling EM wave generates a propulsion force and the vehicle is synchronously driven by the traveling wave. Because of high flux densities, the electrodynamic system offers wide air gaps for levitation, guidance, and propulsion. The levitation and guidance are stable because of the repulsive forces in the right directions.

10.9.3 Programs Undertaken on Levitation Systems

The most important research and development programs on electrodynamic levitation systems were carried out by Germany and Japan in early 1970s. The Germans used a 280-m-diameter circular track to test the 20-ton test vehicle, Erlangen. A cross section of the German test vehicle with a 45° bank angle to minimize the impact of centrifugal forces acting on the vehicle and an end view of a Japanese test vehicle are shown in Figure 10.17. The German test vehicle demonstrated a speed of 150 km/h over a distance more than 1000 km with four superconducting magnets interacting with two rows of aluminum plates installed on the surface of a straight track.

Simens Research Center in 1979 performed levitation tests using a linear synchronous motor (LSM) with 3-phase ac armature windings at the bottom of the track. The LSM generated a thrust of about 4500 lb, accelerating the train to a maximum speed of 150 km/h. Competitive EM-levitation systems were investigated in 1980s with speed capabilities between 300 and 400 km/h. The levitation system program initiated by Japanese National Railway (JNR) carried out successful tests around 1975. The tests demonstrated a speed capability of 517 km/h over a 7-km linear track using a 13-m-long, 10-ton test vehicle. This levitated train was equipped with 8 superconducting magnets (4 magnets for levitation and 4 for propulsion with an LMS and guidance). Cooling of the high-power superconducting magnets was accomplished by sealed helium-bath cryostats refilled between the test runs. Later, the JNR demonstrated the capability of another levitation system using a test vehicle consisting of three single cars over a 25-km linear track with a speed exceeding 520 km/h. The JNR introduced the bullet trains around 1985 between the imperial cities with speeds exceeding 300 km/h, based on superconducting electrodynamic levitation technique. In March 1997, JNR introduced the fastest train in the world, with top speed capability exceeding 550 km/h. Japan has introduced several state-of-the-art superconducting high-speed trains to provide fast, efficient, and reliable transportation to its citizens. Target characteristics for Japanese high-speed trains are summarized in Table 10.5.

10.9 SUPERCONDUCTING ELECTRODYNAMIC LEVITATION SYSTEMS

(a)

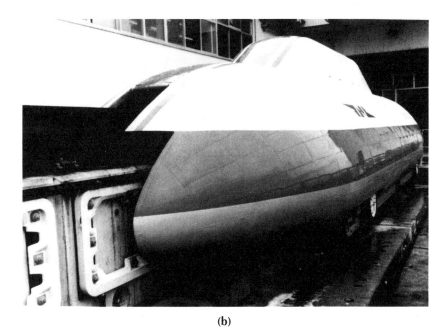

(b)

FIGURE 10.17 Test vehicles and tracks developed by German and Japanese companies: (a) Erlangen test carrier with active track; (b) test vehicle of Japanese National Railways (JNR) on test tract.

TABLE 10.5 Target Characteristics for the Japanese High-Speed Train

Maximum number of cars for one train	16
Maximum speed	550 km/h
Maximum acceleration and deceleration	
Acceleration	3 km/h/s
Deceleration	Normal brake, 5 km h/s; emergency brake, 10 km/h/s
Liftoff speed	Below 100 km/h
Effective levitation height (between coil centers)	250 mm
Track tolerance	±10 mm/10 m
Hours of operation	From 6:00 to 24:00 Intervals—15 min
Time of continuous operation without supply	18 h
Number of superconducting magnets	8 per car
Heat load	50 W/car at 4.2 K
Weight of one car	33–30 t
Body	12 t
Superconducting magnets and cryostats	8–6 t
Refrigerator or helium recovery	3–2 t
Other miscellaneous equipment	4 t
Passengers	6 t

10.10 SUPERCONDUCTING TRANSMISSION LINES AND CABLES

The use of superconductors to transmit electric power has been a subject of great interest for more than two decades [9]. In 1977, Philadelphia Electric investigated the underground transmission of 10,000 MW of electric power over a distance of 106 km. Sixteen transmission systems were considered, including 1 dc and 2 ac superconducting power transmission lines with one overhead transmission line for comparison purposes. The results of that study revealed that the ac superconducting cable operating at 4.2 K results in a transmission with the least cost of all the underground systems. The results further indicated that the dc superconducting high-power cable ranked tenth in terms of cost. The cost of the converters amounted to about 60% of the total cost, which means that the dc superconducting cable operating at 4.2 K will be most attractive for long transmission lines.

The emergence of the new high-temperature superconductors (HTSC) technology has renewed further interest in the application of superconducting cables because of less stringent refrigeration requirements. The lower cost of a refrigeration system at 77 K can make the superconducting cable economically more competitive or even superior at a lower power transmission level of around 1000 MW.

10.10.1 Design Aspects

Superconducting cable [9] requires cryogenic cooling for the entire length of the transmission line. The economics of operation requires that the refrigerant (helium or nitrogen) be returned to the refrigeration system for recooling without any loss of refrigerant. The cooling system must use countercurrent streams within the same cryogenic enclosures for optimum economy. Furthermore, a superconducting cable with high surface-to-volume ratio is essential for high cooling efficiency. In a superconducting cable with a power rating in excess of 5000 MW, the cable must be drawn into a cryogenic enclosure in which the "GO" refrigerant flows through the annulus space outside the cable but within the cryogenic enclosure. The cryogenic enclosure consists of an inner tube surrounded by multilayer thermal insulation and a vacuum jacket. For optimum cable performance, refrigerator spacing from 10 to 50 km [9] must be considered. However, the actual spacing must be determined by taking into account several factors affecting the system optimization: the cost of refrigerators, the cost of cryogenic enclosure, and the acceptable temperature rise of the refrigerant. The size of the cable is strictly dependent on the current-carrying capacity of the superconductor, stabilizing requirements, and amount of electrical insulation involved. The latest superconducting cables use Nb_3Sn filaments in a copper matrix, which offers both the low cost and high transmission efficiency.

Cross sections of various dc superconducting cables with and without heat shields are illustrated in Figure 10.18. In ac superconducting cables, large cable size and cryogenic enclosure are required, which increases the heat leak and the enclosure cost and leads to higher power consumption by the refrigerators. These factors make the fabrication and operation of ac superconducting cables more expensive than dc superconducting cables. In dc superconducting cables, the costs of the converters required at the ends of the transmission line sections amounts to about 60%, according to the Philadelphia Electric Company study [9]. However, the converter cost remains the same irrespective of the transmission line length. The study concluded that lower line cost will eventually make the dc superconducting cables most attractive for cables shorter than 200 km in length.

10.10.2 Cooling Systems

Early superconducting cable designs used metallic, low-temperature superconductors with helium cooling at 4.2 K. The latest high-temperature ceramic superconductors with critical temperatures exceeding 125 K offer minimum cooling system cost at 77-K operation. Operating temperatures at 77 K or higher will allow the use of cheap liquid hydrogen rather than expensive liquid helium or nitrogen. In fact, the cost of cryogenic refrigeration is more dependent on operating temperature level than on the working substance or refrigerant used. Cooling at 77 K is both easy and inexpensive using a bath of liquid

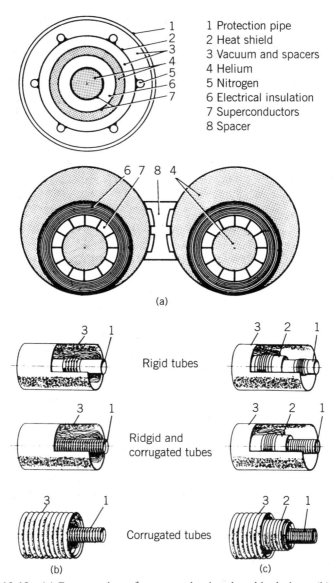

FIGURE 10.18 (a) Cross section of superconducting dc cable designs; (b) cables without cooled intermediate shields, only multilayer superinsulation, and (c) cables with N_2-cooled intermediate shields and multilayer superinsulation: 1, helium tube; 2, N_2 shield; 3, ambient temperature tube. Instead of multilayer superinsulation between N_2 shield and ambient temperature tube, powder fillings can be used.

nitrogen. However, cooling over long distances with high transmission efficiency is possible only with a 4.2-K cooling system.

Refrigerator cooling efficiency is strictly dependent on the thermal insulation known as the cryogenic envelope of the superconducting cable. Thermal insulation is absolutely necessary to reduce the heat flow from outside to the helium-cooled superconducting cable core. Two types of thermal insulation are used (Figure 10.18): (1) flexible cryogenic envelope (FCE), using corrugated thin tubes, and (2) rigid cryogenic envelope (RCE), using rigid tubes. FCE requires no compensation for cold contraction of the helium tubes and can be wound on drums, which is most suitable for long transportation distances. RCE cable construction suffers from greater thermal losses because of the large number of mechanical supports with large surface areas required. The estimated thermal losses (P_{th}) for both types of superconducting cable construction are as follows:

RCE Construction:

$$P_{th} = 100\,\text{mW/cm}^2 \quad \text{for 4–6 K operation}$$
$$= 2\,\text{W/cm}^2 \quad \text{for 77 K operation}$$

FCE Construction:

$$P_{th} = 500\,\text{mW/cm}^2 \quad \text{for 4–6 K operation}$$
$$= 4\,\text{W/cm}^2 \quad \text{for 77 K operation}$$

10.11 SUPERCONDUCTING COMPONENTS FOR ELECTRICAL HIGH-POWER SYSTEMS

This section presents some superconducting components for possible applications to high-power rf and electrical systems. These components include the superconducting fault current limiter (FCL), the superconducting current transfer switch (SCTS) or current diverter, and the superconducting bearing.

10.11.1 Superconducting Fault Current Limiter

The superconducting FCL uses HTSC component technology to provide needed protection for high-power military radar transmitters, such as traveling-wave tubes, cross-field amplifiers, and klystrons. These devices (FCLs) provide adequate protection to costly subsystems or systems from short circuiting or an unexpected surge in the electrical power source. The FCL provides continuous, reliable operation, particularly, in critical and dangerous military environments. Three distinct types of FCLs [8] (Figure 10.19), each rated with specific displacement, offers protection against sudden short circuit-

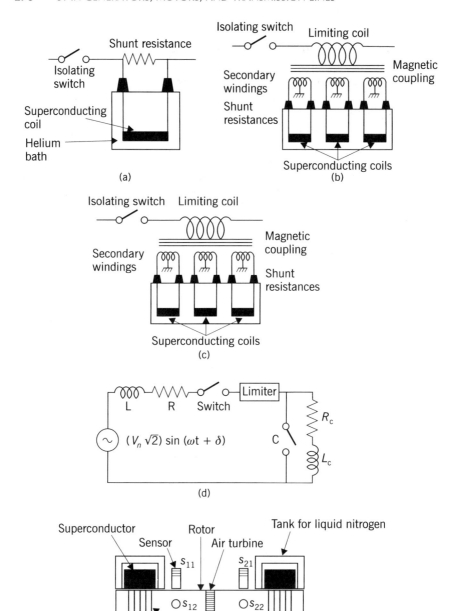

FIGURE 10.19 Superconducting fault current limiters: (a) FCL using resistance; (b) FCL using inductance; (c) FCL using hybrid design [9]; (d) equivalent electrical circuit for FCL where V_n = normal voltage, ω = angular frequency, and δ = voltage angle at the short circuit; and (e) Schematic diagram for a superconducting bearing [8]: S_{11}, S_{12}, S_{21}, and S_{22} are sensors for displacement data.

ing and unexpected surges in the electrical system. These devices also avoid replacement of costly system components damaged by system malfunction. Illinois Superconducting developed this device in 1996 for the U.S. Army Research Laboratory. Major FCL development programs are described in Table 10.6. The device operation is based on the ultrafast transition from the superconducting (nonresistance) state to the normal (resistive) state by overstepping the critical current density. The current limiter clips the electromechanical stresses on the network and the need to handle excessive fault curent [8]. The limiter also provides significant reduction in the fault duration, which should increase the power transmission capability and improve the dynamic stability of the high-voltage system. In essence, the device guarantees the absence of a current excursion higher than a safe predetermined value without the need of a fault-detection system. It is self-operating and repetitive and requires no major maintenance.

Three types of FCL devices are available: resistive, inductive, and hybrid. The hybrid FCL configuration offers the most reliable operation, a significant reduction in cryogenic losses, and lower operating costs. The peak temperature (T_{peak}) is the key performance parameter of this device and is generally specified around 80 K. As the peak temperature of the superconducting cable exceeds 100 K, the cable heats adiabatically with very high thermal excursion, which could burn out the coil due to excessive power dissipation. An optimum value of the temperature time derivative (dT_{peak}/dt) just after the quench (when $t = 0$) is about 300 K/µs for a NbTi superconductor at 4.2-K operation [8]. The peak temperature near 100 K not only increases thermal and mechanical stresses, but also increases the recovery time as well as the reclosing time of the isolating switch after a fault. In brief, the lower the superconducting coil inductance and its peak temperature, the lower will be the temperature rise during the quench.

10.11.2 Superconducting Current Transfer Device (SCTD)

The SCTD offers transfer of current among parallel superconducting branches in a low-voltage dc (LVDC) meshed superconducting system. A superconducting LVDC system reduces transmission and distribution costs. Superconducting transmission lines can connect distant power-generating stations to load distribution systems, replacing high-voltage transmission and substation systems with low-voltage, high-current system. These devices eliminate the need for costly and heavy step-up transformers and significantly reduce the high-voltage insulation requirements. Thus, the complete system can operate at optimum generator voltages, resulting in a single voltage level operation from generation to distribution with minimum cost.

TABLE 10.6 Major Fault-Current-Limiter Development Programs Undertaken by Various Countries

Country	Organization	Type	Specifications			Year Achieved
			Voltage (V)	Current Symmetrical (A rms)	Intercepting Power Rating (kVA)	
Switzerland	ABB	Screened core	480	132–151	100	1993
					1000	*Under way*
Israel	Ben Gurion University	Screened core	1000	25	25	1994
Japan	Criepi	Screened core	50	5	0.25	1994
			6000	*400*	*2400*	*Under way*
Canada	Hydro Quebec	Screened core	450	95	43	1995
					100–200	*Under way*
Germany	Daimler-Benz	Screened core	200	5	1	1995
United States	Lockheed Martin	Electronic/inductive	2400	3000	7200	1995
			15 000	*20 000*	*300 000*	*Under way*
United States	Illinois Superconductor	Resistive	600	100	60	End of 1996
Germany	Siemens AG	Resistive	N.A.	N.A.	0.15	1996
					100	*Under way*
		Low-temperature Superconducting Types				
France	GEC-Alsthom Electric de France Alcatel Alsthom Recherche	Resistive	36 000	210	76 000	1995
			36 000	*1250*	*450 000*	negotiations
United States	Power Superconductor Applications Corp.	Magnetic/inductive	400–38 000	200–1200	200–18 000	Being marketed
Japan	Tokyo Electric Power Co. Toshiba Corp.	Resistive	*6600*	*2000*	*13 200*	*Under way*

Source: IEEE Spectrum, July 1997.
Note: Italics indicate projections.

10.11.3 Superconducting Bearing

Stable levitation systems without physical contacts can be realized through the use of superconducting bearings [9]. The development of a high-temperature superconducting bearing system is based on the strong pinning force. A typical bearing system (Figure 10.19) has a rotor with permanent magnets for levitation and superconductor for support. Axial restoring force and axial displacement, radial restoring force and radial displacement, stiffness of the bearing, repulsive force for a specified displacement, and mass of the rotor are the key superconducting bearing design parameters.

The force–displacement relationship in the axial and radial directions indicates that most of the superconducting bearings work not only as radial bearings, but also as thrust bearings. The bearings are classified according to the displacement around the critical rotor speeds. The three rotors—type 1, type 2, and type 3—are classified according to the displacement around their critical rotation speeds [7]. Their respective displacements are 200, 170, and 620 µm, respectively.

The exponential rotation speed decay of three frictionless, superconducting bearing types can be empirically expressed as

$$N = (42,600)\exp{-\frac{t}{61}} \quad \text{for type 1 rotor}$$
$$= (46,600)\exp{-\frac{t}{56}} \quad \text{for type 2 rotor}$$
$$= (23,200)\exp{-\frac{t}{72}} \quad \text{for type 3 rotor}$$

where N is the rotation speed of the rotor in rpm, and t is the time constant with values anywhere from 50 to 300 s, depending on the axial moment of inertia and damping coefficient of the rotor used. The above equations indicate that type 1 and type 2 rotors rotate stably over the speed range less than 50,000 rpm, whereas type 3 rotors rotate over the speed range less than 25,000 rpm. Each bearing has its own critical speed, depending on its axial moment of inertia. The peak-to-peak displacement of the superconducting bearing with a type 1 rotor is less than 30 µm, except near its critical speed. The current superconducting bearing designs are best suited for industrial applications, where small load capacities are involved.

SUMMARY

Studies performed by various electrical utility companies indicate that implementation of superconducting technology to turbogenerators and transmission lines will significantly improve electrical efficiency and reliability and considerably reduce weight and size. Studies performed on large motors and propulsion systems reveal that integration of low-temperature superconducting technology offers no power dissipation and eliminates iron parts in some parts of the machines. Superconducting fault current limiters offer adequate protection under short-circuit conditions. Superconducting bearings provide longer lives for high-torque motors. HTSC technology is best suited for ship-propulsion systems, electrodynamic levitation trains, and high-power turbogenerators.

REFERENCES

1. C. J. Mole et al. Superconducting electrical machines. *Proc. IEEE* 61:95–105 (1973).
2. Simon Foner, *Superconductor Applications*. Elmsford, NY: Pergamon, 1981, pp. 658–682.
3. T. Takao et al. Statistical estimation of disturbance energy due to conductor motion in rotor windings of superconducting generators. *IEEE Trans. Appl. Superconductivity* 5:361–363 (1995).
4. J. L. Smith et al. Performance of MIT 10 MVA superconducting generator rotor. *IEEE Trans. Appl. Superconductivity* 5:445–449 (1995).
5. Y. Nakagwa et al. Recent development progress of 70-MW class superconducting generators. *IEEE Trans. Appl. Superconductivity* 5:457–460 (1995).
6. J. L. Kirtley, Jr. Large system interaction characteristics of superconducting generators. *Proc. IEEE* 81:449–461 (1993).
7. M. Komori et al. Superconducting bearing systems using high-temperature superconductors. *IEEE Trans. Appl. Superconductivity* 5:634–637 (1995).
8. P. Tixador. Superconducting current limiters: some comparisons and influential parameters. *IEEE Trans. Appl. Superconductivity* 5: 190–197 (1994).
9. J. L. Kirtley. Application of superconductors to motors, generators, and transmission lines. *Proc. IEEE* 77(8):1143–1154 (1989).

CHAPTER ELEVEN

Cryogenic Refrigerator Systems

The cryogenic cooling scheme required to cool a superconducting device or subsystem is an important component of the superconductive system. The size, weight, cost, and complexity of cryogenic hardware depend on the operating temperature and cooling capacity involved. Regular maintenance and servicing of a cooling system are required to ensure safe, steady, and reliable operation of the superconducting system. Commercial applications of superconducting systems indicate that the optimum design of a cryogenic cooler depends on the following criteria:

1. Selection of a cryogenic system design with ease of operation and minimum cost and complexity.
2. Effective integration of the cryogenic equipment with the superconducting device or system.
3. Proper maintenance in the field with minimum cost

Successful integration of these design criteria have been observed in relative few cryogenic systems, but it has been achieved in military infrared surveillance and target acquisition systems, magnetic resonance imaging (MRI) systems, and space reconnaissance sensors.

In this chapter, closed-cycle and open-cycle cryogenic refrigerators or cryocoolers are described with emphasis on cooling capacity, cost, and maintenance issues. A historical review of cryogenic refrigeration system developments is given and survey is presented of commercially available cryocoolers and their applications in various superconducting systems. Performance capabilities of cryogenic coolers are described in terms of operating temperature and cooling capacity. Both high-temperature (77 K) and low-temperature (4.2 K) cryocoolers are evaluated for possible application to high-power generating systems and superconducting motors for ship propulsion and electrodynamic levitated systems.

11.1 CRITICAL REQUIREMENTS AND OPERATIONAL CHARACTERISTICS

Cooling capacity, operating temperature, field maintenance, cooling efficiency, input power requirement, weight, and size are the critical design requirements of a cryogenic cooling system. Optimum performance with minimum cost and complexity should be the principal design objectives. The significance of the operating temperature on these parameters can be seen by comparing the cooling capacity and the weight of various cryocoolers. For example, the specific power requirements of cryocooler with 5-W capacity are 15, 10, 5, and 0.45 kW at 4.2, 10, 20 and 80 K, respectively. The specific weight at these respective temperatures is 200, 40, 15, and 1.5 kg/W. These figures indicate that significant reduction in input power, weight, and size can be achieved at higher cryogenic operating temperatures. In low-temperature cryocoolers, the most benefit can be achieved if the operating temperature can be raised from 4.2 to about 8 K. Thus, an additional order of magnitude of improvement is possible from rapid development of high temperature superconductors. As far as type of cooling is concerned, conductive cooling is most attractive, because it is free from most logistic and reliability problems.

Maintenance cost and interval must be given serious consideration during the selection of a cryocooler for a specific system application. Maintenance interval criterion can be determined only from the data available on the cryocoolers operating in the fields and the experience of the maintenance personnel. Robert Auckermann of General Electric [1] states that a cryocooler operating below 5 K requires more elaborate and more frequent maintenance than a cryocooler operating at a higher temperature (77 K). This means higher maintenance costs because of elaborate helium purification devices to maintain clean working fluid at all times. Liquid helium cryocoolers run successfully and smoothly in the laboratory under controlled environments, but when liquidifiers are incorporated into a commercial system and fielded for application such as MRI, they fail to perform with high reliability. Cryocoolers operating at higher temperatures have been more successful in meeting their stated reliability goals and maintenance requirements.

11.2 CAPABILITIES OF COMMERCIALLY AVAILABLE REFRIGERATORS

Operating temperature ranges for various commercial refrigerators are shown in Figure 11.1. For low-temperature applications over 1- to 8-K, three distinct commercial cryocoolers—dilution and magnetic refrigerators, Collins–helium liquefier, and Gifford–McMahan (GE) and GM/JT cryocoolers with Joule–Thompson valves— are widely used in various superconducting systems.

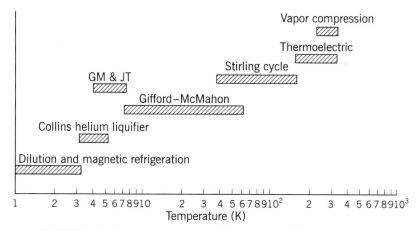

FIGURE 11.1 Temperature ranges for commercial refrigerators [1].

11.2.1 Dilution-Magnetic (DM) Cryocoolers

DM refrigerators are used in very low-temperature applications where a small sample needs to be cooled below 3 K. These cryocoolers are widely used in laboratory environments to perform scientific experiments.

11.2.2 Collins–Helium Liquefier (CHL)

CHLs belong to a broad class of refrigerators based on the Claude cycle operating principle and employ a combination of two or more expansion engines. Recuperative heat exchangers or Joule–Thompson (JT) expansion valves are used to supply liquid at 4.2 K. Specific details on the critical elements of a CHL are illustrated in Figure 11.2. CH refrigerators are commercially available with cooling capacity ranging from 20 to 90 W of refrigeration or 5 to 75 L/h of liquid helium at 4.6-K operation. This particular closed-cycle refrigerator has been widely used in high-energy accelerators, dc superconducting fusion magnets, magnetohydrodynamic (MHD) systems, and magnetically levitated trains. Operating temperatures and potential users of this cryocooler and other commercial refrigerators are shown in Table 11.1. This cryocooler is best suited for a large superconducting system, in which liquid helium from 2.2 to 5 K is used for immersion cooling of the magnets.

11.2.3 Gifford–McMahan (GM) Refrigerators

GM refrigerators have been most successful over the temperature range from 6 to 8 K. They are widely used in cryogenic vacuum pumps and MRI systems to minimize helium boil off through reduction of thermal losses in the cryostat. The GM closed-cycle, two-stage refrigerator shown in Figure 11.3 is most

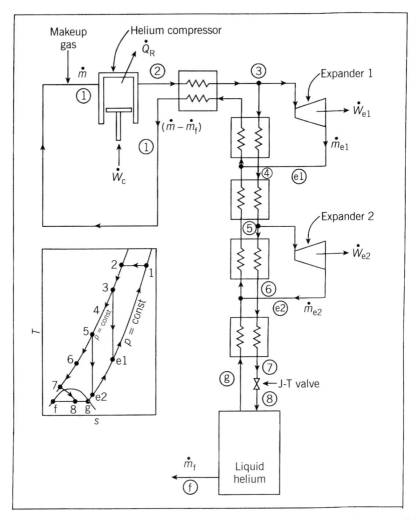

FIGURE 11.2 Schematic diagram for a Collins cycle liquid helium refrigerator [1].

suitable for MRI system application because of minimum loss of helium without adverse impact on operation. There is a tremendous demand for GM refrigerators because of uninterrupted, maintenance-free operation over long duration. It is estimated that more than 1000 units per year are manufactured and distributed worldwide, mostly for MRI systems.

11.2.4 GM/JT Refrigerators

GM refrigerators with JT valves are widely used to provide cooling at 4.2 K. During the period from 1980 to 1995, both the superconducting MRI magnets

11.2 CAPABILITIES OF COMMERCIALLY AVAILABLE REFRIGERATORS

TABLE 11.1 Chronology of Systems Using Closed-Cycle Refrigeration [1]

Application	Temperature (K)	Refrigerator	User
1960s			
Infrared	80	Stirling	U.S. military
Hydrogen bubble chamber	20	GM	High-energy research
Microwave amplifiers	4.5	GM + JT	NASA
1970s			
High-energy accelerators	4.5	Collins	DOE
Dc superconducting motor	4.5	Collins	U.S. Navy
Fusion	4.5	Collins	DOE
1980s			
Fusion magnet test facility	4.2	Collins	DOE
30-MJ SMES	4.5	Collins	BPA
Magnetohydrodynamics	4.5	Collins	Russia
Magnetically levitated train	4.5	Collins	Russia
Magnetic resonance imaging	4.2	GM[a]	Commercial
Cryovacuum pumping	20	GM	Commercial
1990s (potential)			
SQUID	8.5	Stirling	NBS
SSC	2.0	Supercritical	DOE
MRI	10	GM	Commercial
Magnetically levitated train	—	—	DOT
SMES	—	—	Industry
Infrared	65	Stirling	U.S. Military
Infrared	10	Stirling	NASA
High T_c	20–30	—	DOE/military
Infrared	4.2	Boreas cycle	U.S. military

[a] Used for shield cooling only.

and cryopumps made exclusive use of two-stage GM refrigerators to provide conduction cooling instead of immersion cooling. Conduction cooling is generally free from logistic and major reliability problems. In an MRI system, the superconducting magnet is immersed into a liquid helium bath at 4.2 K. The GM cryocooler cools the magnetic radiation shield used to reduce the helium boil off to a very low level, which can be easily maintained with infrequent helium deliveries as seldom as once a year. Cryopumps generally use GM refrigerators to cool the cryopumping surfaces to 20 K.

Accelerated research and development activities on conductively cooled magnets for MRI applications and high-temperature superconducting materials will improve the performance level of MRI systems with minimum cost and improved reliability. An MRI conductively cooled magnet uses a two-stage GM refrigerator to cool both the magnet and the radiation shield, thereby requiring only a few watts of refrigerant. SQUID technology for military,

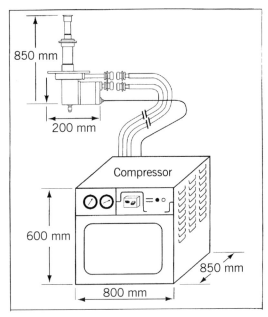

FIGURE 11.3 Gifford–McMahan cycle, two-stage refrigerator for MRI applications [1].

medical, and scientific applications is progressing rapidly and can be integrated easily with closed-cycle GM or Stirling cryocooler designs in near future.

From 1960 to 1980, GM/JT cryocoolers were widely used in low-noise microwave amplifiers and the Stirling coolers were used in a number of military IR systems, demonstrating a maintenance interval exceeding 3000 h. Both MRI and cryopumping require two-stage cooling, one over 80 to 20 K with 40 to 50 W of cooling requirements and the other (second-stage cooling) at 4.2 K with 4 W of cooling power. These cryocoolers need only yearly maintenance, because of low helium loss and minimum downtime.

11.2.5 Stirling Cryocoolers

Research in bearing and seal technologies and the development of high-quality lubrication have led to an optimum design of a Stirling cooler, which has demonstrated several years of unattended operation. Its potential applications include space sensors and unattended systems in secluded areas, where continuous operation with high reliability is important. Stirling-cycle (SC) refrigerators are widely used in systems where small size and weight and high reliability over extended periods are the principal requirements. These cryocoolers are best suited for systems requiring small cooling capacities of 1–5 W over 60 to 80 K. SC coolers with unique linear drive mechanisms instead of old rotary drive mechanisms are in great demand for space and military applications, because of improved reliability. The new Stirling cooler design uses an orifice pulse tube, which eliminates the moving displacer from the expander, thereby leaving only one moving component in the compressor. The linear drive mechaism can be operated down to 60 K without any compromise in reliability. Multistage designs of this cooler are in the developmental phase.

11.2.6 Self-regulated Joule–Thompson Cryocooler

This GM cryocooler has Joule–Thompson valves, which allow liquid helium cooling at 4.2 K or lower with minimum loss of the refrigerant without compromising the reliability or overall performance. The self-regulation [2] of the valve offers further improvement of the cryocooler.

11.3 CLOSED-CYCLE CRYOGENIC (CCC) REFRIGERATOR

The CCC refrigerator, sometimes referred as a cryocooler, is viable alternative to a liquid helium cooling system. The CCC uses helium gas as the working fluid. The compact closed-cycle helium gas cryocooler shown in Figure 11.4 is most attractive for SQUID sensor application and has only two moving parts, a piston and a displacer. During the compression mode, heat is produced,

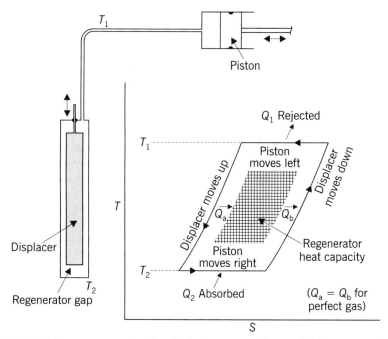

FIGURE 11.4 Compact, closed-cycle, helium gas, split-type Stirling cryocooler.

which is rejected at the ambient temperature T_1. During the expansion mode, the fluid is expanded, producing a flow of heat into the field at temperature T_2. The fluid picks up the heat from the walls of the annular gap as it travels from T_2 to T_1. For a steady-state operation the heat Q_a must be equal to Q_b.

The enthalpy change in heat quantities at the two temperatures depends on pressure, and the expansion cannot be isothermal, which is a serious limitation in that the heat capacity of the walls at cold ends becomes insufficient. This drawback can be overcome by using advanced materials with large heat capacity and high thermal conductivity in a regenerative heat exchanger at low temperatures. The material selected must provide adequate protection against electromagnetic interference and noise entering the SQUID sensor.

Specific details on a single-stage displacer refrigerator using advanced nonmagnetic materials are illustrated in Figure 11.5. This cooling machine operates at 1 Hz and attains a temperature of 50 K within less than 4 h, which can be maintained for about 2 days on a cylinder of helium.

Based on single-stage Stirling cryocooler design, a three-stage, closed-cycle split Stirling cryocooler (Figure 11.6) was developed using advanced components, namely, nylon displacer, epoxy–glass cylinder, and aluminized-plastic radiation shield. This machine can maintain an operating temperature below 16 K when operated at 1 Hz. By eliminating the dead space produced by the

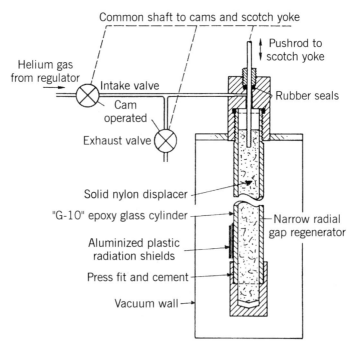

FIGURE 11.5 Single-stage, displacer–regenerator cryocooler using advanced nonmagnetic materials.

11.3 CLOSED-CYCLE CRYOGENIC (CCC) REFRIGERATOR

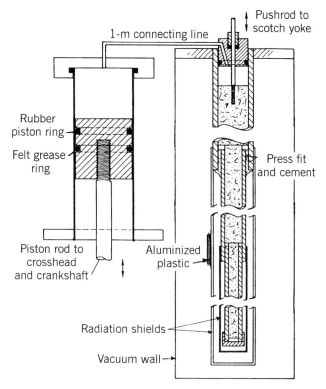

FIGURE 11.6 Three-stage, closed-cycle, split-type Stirling cryocooler using advanced nonmagnetic materials.

difference in coefficients of expansion of various materials used and optimization of the phase angle between the piston and displacer, this cryocooler can maintain a temperature below 13 K over a continuous duration as long as 5 weeks. The cryocooler demonstrated a nonstop operation over 5000 h with no sign of wear and tear, while maintaining a temperature between 12.5 and 13 K at the cold end with no heat load. The outer radiation shield is maintained at 120 K and the inner radiation shield at 40 K. When operated at 1 Hz, the mechanical power required during the compression cycle is about 15 W, assuming an isothermal process. Since a considerable fraction of work is returned during the expansion cycle, the net power required is estimated to be less than 10 W.

This machine is capable of maintaining low temperature over a very long duration with very low power consumption. As stated earlier, a low-power cryocooler can maintain operating temperatures below 15 K with 10 to 20 mw of heat load. This multistage cryocooler distributes the refrigerator capacity in such a way that the bulk of the heat input due to conduction and radiation is pumped at relatively higher temperatures, leaving very little heat for lower temperatures. The optimum distribution of the refrigeration capacity versus temperature is very important in a cryogenic cooler design. A typical small liquid-helium cryostat with an evaporation rate of 1 L/day, which corresponds to 700 L of gas at room temperature, can support a total heat load of 29 mw at 4.2 K and still have the capability to absorb an additional load of 7.6 mw/K due to the heat capacity of the vapor. The cost of liquid helium will increase as the supply diminishes, whereas the cost of cryocoolers will decrease significantly, if produced in quantity.

11.4 COOLING SCHEMES USED BY VARIOUS CRYOCOOLERS

The selection of a cooling scheme is strictly dependent on the cooling capacity requirement and the type of cryogen used. The characteristics of common cryogens are summarized in Table 11.2

The choice between immersion cooling and conduction cooling is a fundamental design issue for a cryocooler. For immersion cooling using liquid helium, one can use a Collins liquefier for a closed-cycle operation or for a open-cycle operation with supply of liquid helium as needed. For low thermal loss systems on the order of 5 W or less, GM/JT cryocoolers are widely used. Under transient state operation during cooldown phase, the helium boil off is allowed to escape from the cryostat and the cryostat is provided with new supply of liquid helium. Frequent maintenance is desirable irrespective of cooling scheme or cryogen used to keep the system operating with high efficiency and reliability over extended periods.

A conduction cooling scheme eliminates the need for frequent liquid helium refills, thereby yielding maximum economy. Most commercially available cryocoolers have small cooling capacities of less than 5 W and provide a single point of cooling. The use of such cryocoolers requires complete familiarity with thermal load requirements, maximum allowable operating temperatures, and optimum thermal design to minimize temperature differential between the superconductor and the cryocooler.

The cryocoolers must be designed to handle heat loads with adequate safety margins during the most critical operation. The transient-state operation

TABLE 11.2 Characteristics of Potential Cryogen

Cryogen	Boiling Temperature (K)	Relative Vaporization	Power (W/W)
Helium	4.2	1[a]	1000
Nitrogen	77.0	64	30
Neon	27.2	41	140

[a] 28 mw boils 1 L of liquid helium per day.

presents the greater thermal loads, which can limit the duty cycle for the systems shown in Table 11.3.

The environmental factors, namely, ambient temperature, shock, and vibrations, and availability of proper maintenance service must be taken into account during the selection of a cryocooler. A sophisticated high-capacity cooling scheme using liquid helium (Figure 11.7) is best suited for ship-propulsion applications, where reliable, uninterrupted operation over long duration is the principal requirement. This cooling system provides safety interlocks to protect the system against high combat temperatures, contamination, insufficient cooling water, and low supply voltage.

11.5 HIGH-TEMPERATURE COOLING SYSTEM FOR SONAR TRANSMITTERS

Recently an acoustic sonar transmitter using high-temperature superconductor technology was developed under the Small Business Innovative Research contract [3]. This superconducting cooling system designed to cool the sonar transmitter illustrated in Figure 11.8 represents a classic example of integrating three distinct technologies—HTSC technology, magnetostrictive technology, and cryocooler technology. This represents the first successful application of superconducting technology to an acoustic transducer without using a liquid cryogen. The system comprises magnetostrictive elements, a pair of magnetic coils wound from HTSC wire, and a Stirling cycle refrigerator to maintain an operating temperature of 50 K. Cooling was maintained under both ac and dc currents in the HTSC coils without liquid cryogen bath, thereby yielding maximum economy and reliability. The cooling of the acoustic transducer must be very reliable and effective, because it requires very high power at low frequencies for accurate mapping and detection over long distances. Potential applications of cryogenically cooled sonar transducer include seismic tomography, global warming measurement, ocean floor mapping, and ocean current monitoring. Because it has few moving parts, the reliability of this cooling system is inherently very high.

TABLE 11.3 Thermal Losses for Small Conductively Cooled Systems

Thermal Loss Type	First Stage	Second Stage
Temperature (K)	40	10
Conduction/radiation (W)	17.0	0.4
Power leads (W)	14.0	2.0
Margin allowed (W)	7.8	0.6
Total thermal loss (W)	38.8	3.0

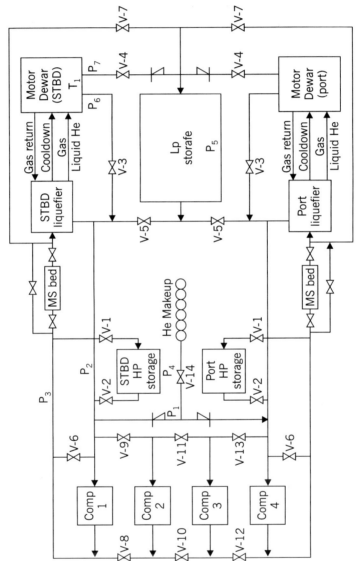

FIGURE 11.7 Liquid helium supply and management system for ship propulsion application [1].

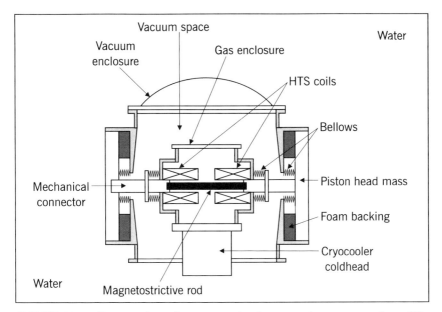

FIGURE 11.8 Cross section of a superconducting acoustic sonar transducer [3].

11.5.1 Cooling Power Levels at High Superconducting Temperatures

The critical current in HTSC wires increases linearly with the decrease in operating temperature, whereas the cooling power required by the cryogenic refrigerator follows an exponential relationship with decreasing temperature. When the operating temperature is decreased from 77 to 50 K, the critical current is doubled, while the cooling power requirement increases by only 15%. Cooling for a high-power system can be provided by a Stirling-cycle cryocooler designed for cryogenic electronics applications with a typical capacity of 250 W at 77 K. This cryocooler is best suited for a sonar transducer operated at 50 to 55 K with a cooldown time of less than 4 h. This cooler is acoustically quiet and has an impressive reliability of better than 50,000 h. The cooler can operate successfully in all positions and does not require periodic replenishment of liquid cryogen. This cooler is best suited for ship-propulsion superconducting systems, superconducting motors and generators, magnetic storage devices, and low-cost, portable MRI equipment.

11.6 PROGRESS IN MINIATURIZED CRYOCOOLERS WITH HIGH RELIABILITY

Miniaturized cryocoolers have potential applications to space sensors and sophisticated military infrared systems such as IR line scanners, IR search and tracking systems, thermal imaging radiometers, and IR missiles. The design of a microcooler uses the Stirling-cycle engine principle to generate the liquid nitrogen temperature (77 K) in less than 3 min. The microcooler incorporates a regenerator to create a compression–expansion refrigerator system with no valves. The regenerator has a high heat capacity, and acts as a highly efficient heat exchanger. The electronic sensing devices control the cooling action to maintain the required 77-K operating temperature. A microcooler designed for NASA thermal imaging system weighs only 15 oz, consumes electrical power of less than 3 W, has a shelf life of 5 years, and has demonstrated continuous operation exceeding 8000 h. High reliability is due to the selection of improved materials with self-lubrication features, very low-friction clearance seals, elimination of gaseous contamination, and use of linear motor drive. The 1-W and $\frac{1}{4}$-W split-Stirling coolers with linear motor drives are capable of providing continuous operation over more than 7000 h. A microminiaturized cryocooler provides a cooling capacity of 150 mw at 77 K with input power less than 3.5 W with a 150-mw heat load and cooldown time ranging from 1.5 min at 120 K to 4.0 min at 77 K from a room temperature of 25°C. The input electrical power requirements for a microcooler as a function of heat load and operating temperature are shown in Figure 11.9.

11.7 CRYOCOOLER DESIGN USING RARE-EARTH ELEMENTS AS REGENERATOR MATERIALS

The first 4.2-K, two-stage GM/RE [4] cooler used rare-earth materials as regenerator elements. GM/RE machines can reach 4.2 K in a short time, because rare-earth material such as erbium nickel has a significant heat capacity below 10 K. The normal capacity of GM/RE cryocooler is about 100 mw at 4.2 K, but a capacity of 1.05 W has been demonstrated using a unique combination of rare-earth materials.

11.8 CRYOCOOLER WITH HIGH-PRESSURE RATIO AND COUNTERFLOW DESIGN

Studies performed on cryocooler designs indicate that the use of a high-pressure expansion ratio and counterflow heat exchanger are essential to achieve a low-cost, high-efficiency cryocooler with improved reliability. The revolutionary design of a Boreas cryocooler uses both the high-pressure ratio and counterflow heat exchanger, thereby eliminating the need for costly high heat capacity rare-earth regenerator materials to achieve cooling at 4.2 K. The Boreas cycle uses a wet expansion engine to produce liquid helium.

The three-state cryocooler illustrated in Figure 11.10 provides a regenerative heat exchanger in the two warm stages at 70 and 20 K and counterflow heat

11.8 CRYOCOOLER WITH HIGH-PRESSURE RATIO AND COUNTERFLOW DESIGN

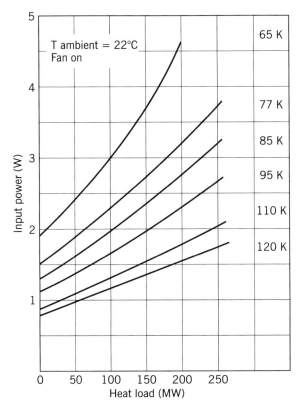

FIGURE 11.9 Input power requirements for a microcooler as a function of heat load and temperature [4].

exchange in the third stage to reach to 4.2 K. This cryocooler integratres regenerative and counterflow heat exchangers with a three-stage expander to yield a compact cold head with a single helium flow circuit, thereby resulting in a simple, reliable, inexpensive cryocooler design.

A pressure–volume (P–V) schematic diagram and the 4.2-K stage efficiency as a function of load temperature are shown in Figure 11.11. It is evident from these curves that the cold-stage efficiency improves with increasing pressure. The expansion cycle reduces the gas temperature. Helium vaporized by heat loading is returned to the compressor via the counterflow heat exchanger, thereby recuperatively cooling the cylinder and displacer in preparation of the next operation cycle.

11.8.1 Thermodynamic Aspects of the Boreas Cycle

A rigorous thermodynamic comparison between the GM and Boreas cycles for 4.2-K cooling reveals that the Boreas cycle is four times more efficient than the

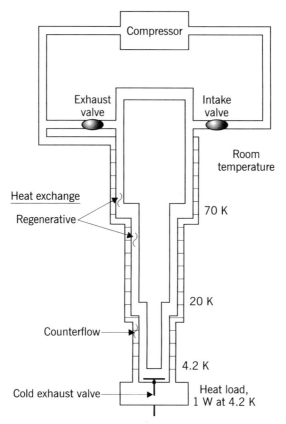

FIGURE 11.10 Improved design of a 3-stage Boreas cryocooler using advanced technologies.

GM cycle. The comparison further indicates that the efficiency of the Boreas cycle is a strong function of the compressor suction pressure. The efficiency curve of Figure 11.11 shows that the Boreas cycle is more efficient than the GM cycle at refrigeration temperatures below 7 K. One can see from the efficiency curve that a 4.2-K operation, the Boreas cycle offers efficiency greater than 70% with a suction pressure of 20:1 atm, while the GM cycle typically operating with a suction pressure of 20:6 atm can achieve an efficiency of only 17%. The high efficiency of the Boreas cycle is strictly due to the high-pressure expansion ratio to achieve 4.2 K. Furthermore, the high-pressure ratio is feasible only with a counterflow heat exchanger.

11.8.2 Thermodynamic Efficiency Comparison for Various Coolers

Thermodynamic efficiency for the 1-W, 4.2-K cryocooler can be evaluated by looking at the power consumption relative to the ideal Carnot cycle.

11.8 CRYOCOOLER WITH HIGH-PRESSURE RATIO AND COUNTERFLOW DESIGN

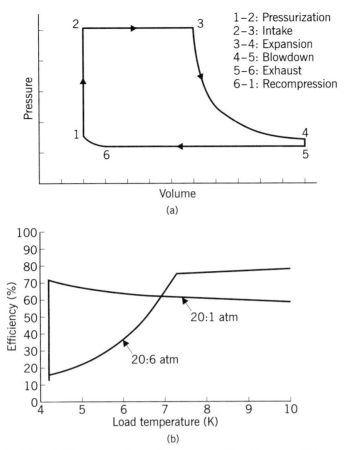

FIGURE 11.11 (a) Pressure–volume diagram and (b) cold-stage efficiency of a 3-stage Boreas cryocooler [4].

Theoretically, the input power required per watt refrigeration at 4.2 K is 71 W. A commercially available GM/JT cryocooler, which requires 4.5 kW input power to produce 1 W at 4.2 K, has an ideal efficiency of only 1.6%. A GM/RE (rare-earth) cryocooler that uses 0.524 kg of multiple rare-earth materials requires 7 kW of input power to produce 1.05 W at 4.2 K and operates at 1.1% efficiency of Carnot cycle.

11.8.3 Advantages of High-Pressure Ratio Expansion

The high-pressure expansion ratio offers significantly improved efficiency, minimum power consumption, lower operating costs, reduced helium mass flow, lower cold head speed (22 rpm) for Boreas cycle versus 70 to 140 rpm for GM cryocoolers), high reliability due to minimum wear/tear of parts, longer compressor life, and practically vibration-free operation. Quiet, low-vibration, cold head operation is highly desirable in many applications, because of the sensitive electronic components used. In medical applications such as MRI, patient comfort and low magnetic noise level are additional requirements. Boreas cryocoolers offer these advantages because of their slow speed and operation with minimum vibrations. Coolers operating with high-pressure ratios open new opportunities in the superconducting market for future cryocoolers, and will demand higher efficiency, lower power consumption, lower operating cost, higher reliability, and virtually vibration-free operation.

11.9 OPTIMIZATION OF THE COOLING CAPACITY OF A CRYOCOOLER

Maximum available cooling capacity of a cryocooler is the most important performance parameter. Discussion on this subject is limited to the split-type free-displacer (STFD), Stirling refrigerator system illustrated in Figure 11.12, because of its several advantages. Maximum available cooling capacity (MACE) is a performance indicator of a cryocooler and must be optimized for a specific application. The net cooling capacity (Q_{net}) as a function of cold-end temperature (T_L), normalized displacer natural frequency (f_n/f), and

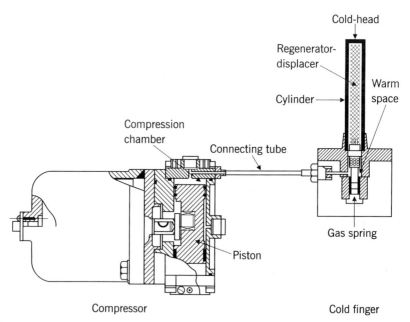

FIGURE 11.12 Critical elements of a Stirling refrigerator design with optimum cooling capacity [5].

11.9 OPTIMIZATION OF THE COOLING CAPACITY OF A CRYOCOOLER

displacer loss coefficient (C_d) for a split-type Stirling cryocooler is shown in Figure 11.13. T_W and T_C indicate warm gas temperature and compressor temperature, respectively.

The MACE of an STFD Stirling refrigerator, which is defined as Q_{min}, can be calculated by integrating the expansion space represented by the pressure-volume curve. In brief, the quantity Q_{min} is proportional to the integral of expansion pressure and the displacer displacement as shown at the top of Figure 11.11. The net cooling capacity is equal to the maximum cooling capacity less the heat conduction loss of regenerator, enthalpy flow loss of regenerator, shuttle heat loss of displacer, and hysteresis loss of the gas.

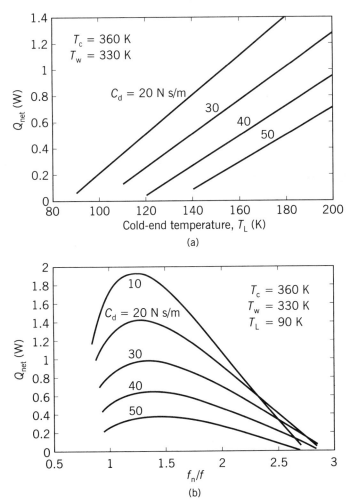

FIGURE 11.13 Net cooling capacity variation as a function of (a) cold-end temperature and (b) frequency [5].

Cryocooler design analysis performed by B. J. Huang et al. [5] using linear network modeling indicates that the net cooling capacity (Q_{nwt}) increases with increasing cold-end temperature and decreasing displacer loss coefficient (Figure 11.13). The linear model further indicates that for a given C_d, the net cooling capacity first increases with increasing frequency, reaches a peak, and then decreases. An optimum value of the net cooling capacity occurs when the normalized natural frequency is between 1.2 and 1.5, depending on the value of the coefficient C_d (Figure 11.13). The value of the loss coefficient depends on the displacer seal design, material used, and the quality of the workmanship. Improved accuracy in the performance prediction of a cryocooler is possible through the modified linear network analysis described in Reference [5]. The loss coefficient must also take into account the frictional and gas leakage losses of the displacer. This means that the loss coefficient C_d is not constant, but can vary as a function of displacer oscillating frequency (f) and cold-head temperature (T_L). An empirical correlation of C_d can be obtained in terms of these parameters. The net cooling capacity as a function of correlated value along with an assumed value of the displacer loss coefficient is illustrated in Figure 11.14. The upper curve in Figure 11.14 yields a more accurate estimate of the net cooling capacity of the cryocooler as a function of cold-end temperature.

11.10 TEMPERATURE STABILITY AND OPTIMIZATION OF MASS FLOW RATE

Transient behavior of a cryocooler is important, particularly when the cooler is used in a critical space or complex military system. The transient behavior of a self-regulating cryocooler can be accurately predicted by the numerical simula-

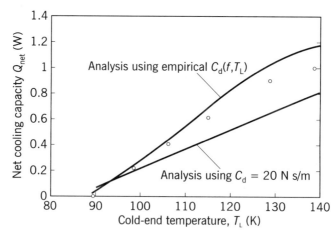

FIGURE 11.14 Net cooling capacity as a function of cold-end temperature and loss coefficient obtained using two different methods [5]: $T_c = 360$ K, $T_w = 330$ K; O, test results.

tion proposed by S. B. Chien [2] which involves modeling various types of bellows control mechanism for a self-regulating cryocooler (Figure 11.15). A flow-regulating mechanism in a cryocooler offers long-term, low gas consumption operation with maximum economy. A bellow control mechanism is widely used for flow regulation. The temperature-sensitive bellow senses the temperature at the cold end and regulates the opening of the JT cooler orifice, thereby allowing adequate gas flow to maintain a specified cold temperature and to minimize the excess flow to avoid waste of the refrigerant.

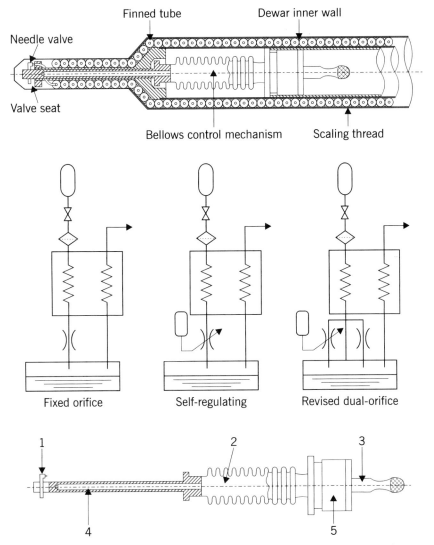

FIGURE 11.15 Bellows control mechanism for a self-regulating Joule–Thompson cryocooler [6]: 1, valve stem; 2, bellows; 3, gas filling tube; 4, sensing tube; 5, support set.

Transient simulation data as a function of various temperatures, mass flow rate, orifice opening, and nose area ratio (NAR) are shown in Figure 11.16. T_h, T_g, T_l, and T_{n2m} represent high-pressure temperature, glass Dewar temperature, low-pressure temperature, and mean temperature of the nitrogen gas, respectively. The parameter m stands for mass flow rate of the gas expressed in kilograms per second. The temperature-sensitive bellow mechanism allows a high flow rate to meet fast cooling requirements and throttles down to conserve

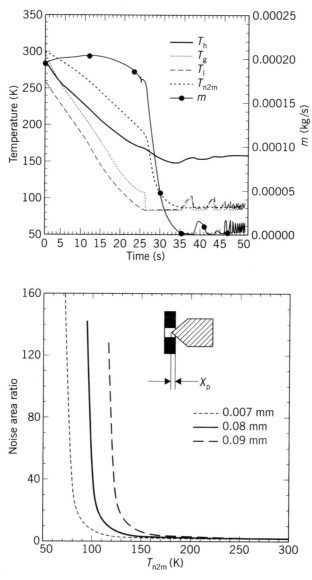

FIGURE 11.16 Transient simulation as a function of temperature and mass flow rate [6].

11.10 TEMPERATURE STABILITY AND OPTIMIZATION OF MASS FLOW RATE

the cooling gas. The structure of the needle valve provides variation of valve opening, which regulates the gas flow rate. The nose area ratio parameter as a function of mean temperature of the gas inside the bellows indicates the level of the valve opening, which is defined as the area ratio of the maximum valve opening to a specific valve opening condition. The spatial temperature variations for high-pressure gas and the glass Dewar as a function of assembly position with respect to initial position for several progressive time increments are shown in Figure 11.17. According to the curves in Figure 11.17, it will take about 30 s for the temperature to drop from 300 to 80 K.

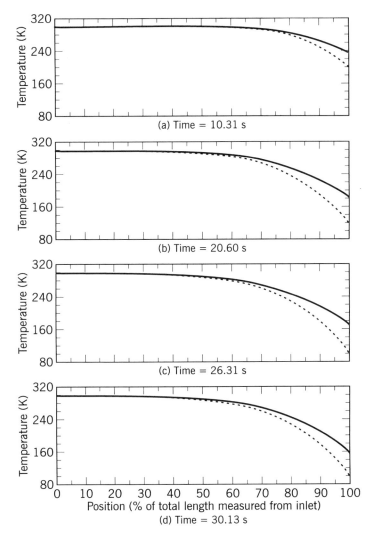

FIGURE 11.17 Typical spatial temperature variations for a cryocooler with progressive time increments [6]: —, T_h, ..., T_g.

SUMMARY

Superconducting devices or systems cannot operate without refrigerators or cryocoolers with appropriate cooling capacities. A cryogenic cooling subsystem is an important and costly component of superconducting system. Weight, size, cost, and complexity of cryocooler hardware depend on the operating temperature and cooling capacity requirements. Regular maintenance and servicing of a cooling system are required to ensure safe, steady, and reliable operation of a superconducting system. Selecting a cryocooler for a specific application requires careful evaluation of cost, reliability, complexity, field maintenance, and frequency of refrigerant refills. Input power requirements of a cryocooler are a function of cooling capacity and operating temperature. Microcoolers are widely used in superconducting airborne military equipment, space sensors, and satellite reconnaissance systems, where weight, size, extended operation, and reliability are the principal requirements.

REFERENCES

1. R. A. Auckermann. Closed-cycle refrigeration for superconducting applications. *Superconducting Industry*, 15–24 (Fall 1993).
2. S. B. Chien et al. A study of transient characteristics of a self-regulating Joule–Thompson cryocooler. *Cryogenics* 36:979–984 (1996).
3. C. H. Joshi et al. Putting a chill into the HTSC applications. *Superconducting Industry*, 26–29 (Fall 1993).
4. M. Wilson. Advances in 4.2-K cryocooler technology. *Superconducting Industry*, 30–35 (Fall 1993).
5. B. J. Huang et al. Split-type free-displacer, Stirling refrigerator design using linear network analysis. *Cryogenics* 36:1005–1017 (1996).

Index

Acoustic:
  saw, 60
  transducer, 17, 285
Amplifier:
  DH-HEMT, 148
  feedback, 147
  gain, 142
  high power, 125
  low-noise, 26
  MESFET, 130
  mm-wave, 125
  monolithic, 145
  noise figure, 142, 143
  nondegenerative, 150
  operational, 149
  parametric, 171
  PARAMP, 150
  solid state, 142
  ultra-low noise, 172
Analog-to-digital (ADC) converter:
  ADC, 26, 80, 177
  architecture, 179, 180
  elements, 173
  high-resolution unit, 182
  quantization, 177
  sigma-delta, 181
  unit, 173
Antenna:
  dipole, 82
  efficiency, 82
  electrically small, 82
  electronically-steerable, 92, 153
  feed-assembly, 186
  helical, 84
  log-periodic array, 84
  loop dipole, 84, 155
  patch, 26
  planar, 209
  printed-circuit, 69
  sidelobe level, 94
  spiral, 84
  superconductor, 82
  transmit/receive, 153
Atomic:
  orbitals, 4
  structures, 7, 15

BCS theory, 1
Bellow:
  control mechanism, 295
  temperature-sensitive, 295
Biomagnetic:
  brain/nerves/eyes, 214
  heart activity, 233
  measurements, 214, 217
  signals, 216
  sources, 217
Biomedical:
  application, 214
  scientists, 219
Bolometer:
  conventional, 196
  hot, 189
  hot electron, 198
  microbolometer, 198
  superconducting, 196
Boreas:
  cryocooler, 288, 290
  cycle, 288, 289
Brain:
  signals, 230
  surgeons, 231

Carriers:
    photon-generated, 190
    thermally-generated, 190
Cardiograms:
    electro, 215
    magneto, 215
Cardiologists, 231
Chemical reaction, 5, 24
Circuits:
    active, 60, 88
    bipolar, 126
    digital, 178
    electronic, 178, 183
    helix, 170
    high speed logic, 136
    integrated, 125
    logic, 18
    microelectronic, 58
    mm-wave, 128, 131
    optoelectric, 125
    passive, 26, 60, 68
    RSFQ, 178
    sampler, 178
Clinical:
    aspects, 214
    assessment, 230
    community, 231
    researcher, 217, 224
Clutter:
    environment, 185
    severe, 172
    signal, 173
Coil:
    feedback, 218
    input, 217
    noise cancellation, 224
    pickup, 217
    signal, 217
Communication:
    infrared (IR), 212
    ship-to-ship, 212
    systems, 82, 88, 124, 184, 212
Components:
    active, 26, 60, 88
    ADCs, 26, 80
    delay lines, 26, 69, 76
    electro-optical, 116
    filter, 26
    optical, 26
    passive, 26, 60, 68
    phase shifters, 26, 88
    radio frequency (rf), 142
    solid state, 26
    SQUID, 26

Compounds:
    bismuth-based, 3
    ceramic-based, 5
    mercury-based, 58
    thallium-based, 3
    yttrium-based, 3
Computer tomography (CT), 214
Conduction:
    cooling, 7, 31
    path, 11
Constant:
    anisotropic, 108
    attenuation, 66
    phase, 68
    propagation, 66, 98
    thermal time, 114
Cooling:
    capacity, 80, 275–276
    conductive, 276
    detector, 80
    efficiency, 269, 276
    immersion, 277
    scheme, 284
    system, 267
Coolers:
    Boreas cycle, 80
    multistage, 281
    Stirling cycle, 80
Coupling:
    mechanism, 12
    strength, 22
Critical:
    current density, 12
    diffusion process, 22
    magnetic field, 15, 22, 33
    rotating speed, 273
    temperature (T_c), 14
Cryocooler:
    closed-cycle, 191
    cost, 177
    design, 177
    efficiency, 177
    GM, 279
    GM/JT, 276
    heat capacity, 186
    liquid-helium, 276
    miniaturized, 287
    reliability, 177
    self-regulation, 281
    single stage, 282
    Stirling, 280
Cryogen:
    characteristics, 284
    liquid, 285

Cryogenic:
  closed-cycle, 281
  cooler, 275
  cooling, 173, 185, 267, 276
  cooling scheme, 247
  enclosure, 212, 267
  envelope, 269
  hardware, 275
  refrigeration, 267
  refrigerators, 275
  regenerator, 282
  self-regulation, 294
  temperature, 67, 103, 106
  vacuum pump, 272
  windows, 110
Cryogenically-cooled:
  amplifiers, 116
  frequency multipliers, 16, 163
  microwave receivers, 172
  mixers, 111, 164
  oscillators, 111
  rf sources, 166
  windows, 111
Cryopump, 279–280
Cryostat, 284
Crystal:
  co-doped, 202
  laser, 204
  lattice, 2
  rare-earth, 202
Current density:
  critical, 12
  transport, 22
Curtic model, 131

Detectors:
  bolometer, 80
  detectivity, 81
  IR, 18, 69, 80
  optical, 20, 189
  photon, 80, 189
  photovoltaic, 191
  quantum, 190
  response, 195
  responsivity, 86, 197
  SIS, 193
  ternary alloy, 19
Devices:
  analog, 20
  analog-to-digital (A/D), 26
  charge-coupled, 205
  charge-injection, 205
  charge-transfer, 205
  data storage, 1, 21
  digital, 8
  solid state, 116
  SQUID, 26
  two-terminal, 116
  three-terminal, 116
  vertical, 147
Diagnostic:
  capabilities, 214
  clinical services, 235
  devices, 214
  equipment, 214
  tools, 214
Diamagnetism, 1
Diffusion effect, 1
Diodes:
  GUNN, 116, 122
  IMPATT, 123
  LED, 210
  mean-time-between-failure (MTBF), 122
  microwave, 116
  PIN, 116
  quantum-well (QW), 116
  reliability, 123
  varactor, 116
Dopants, 7
Dynamic range, 176, 180, 190, 218

Efficiency:
  antenna, 82
  coupling, 153
  drain, 131
  electronic, 111
  external, 200
  feed, 82
  heat transfer, 120, 170
  injection, 126
  internal quantum, 200
  power-added, 125, 138
  radiation, 82
  total differential quantum, 201
Electrical:
  doping, 4
  generator, 20, 23
  motor, 23
Electromagnetic (EM):
  interference, 282
  signals, 178, 195
  torque, 282
Electromechanical stresses, 271
Electronics:
  beam, 94
  cancellation, 231
  charge, 177

302 INDEX

Electronics (*Continued*)
   circuit, 178
   scanning, 92
Electrons:
   cyclotron, 111
   density, 138
   excited, 199
   free, 150
   heating, 198
   hot, 198
   mobility, 125, 131, 136, 142
   normal, 35
   superconducting, 35
   velocity, 131
Electrooptical (EO) components, 189
Energy:
   gap, 1, 9
   pinning, 14
   resolution, 217
   thermal, 4
   white noise, 2, 21
Environment:
   clutter, 172
   jamming, 172

Feedback:
   bias control, 226
   positive, 226
Ferrite:
   circulators, 104
   devices, 88
   isolators, 104
   linewidth, 102
   nonlinear effects, 102
   permeability, 91, 107
   phase shift, 89
   toroid, 98
Fiber-optic (FO) ring laser, 190
Films:
   epitaxial, 100
   thick, 20, 58, 159
   thin, 9, 58, 159
Filters:
   bandpass, 69–70
   band-reject, 86
   CHIRP, 77, 174
   combline, 70
   delay line, 174
   dielectric, 159
   diplexing, 68
   E-plane, 102
   low-pass, 72
   matched, 77
   microwave, 159
   MM-wave, 73
   stripline, 69

   transversal, 77
   tunable, 88
   YIG-tuned, 88
Flux:
   density, 216
   expulsion, 21
   flowmeters, 114
   linkage, 216
   magnetic, 14, 18
   modulation, 231
   noise, 231
   noise level, 216
   pinning, 22
   quantum, 93, 177, 182, 216
   transformers, 218, 221, 234
Focal planar array, 81, 204
Frequency:
   bandwidth, 108
   corner, 217
   crossover, 156
   cutoff, 128
   gap, 193
   optical, 172
   precessional, 110
   pump, 152, 171
   resolution, 176
   resonance, 100
   shift, 96
   signal, 171
   stability, 122, 166
   transit, 128
   unity gain, 136

Gain:
   exchangable, 152
   maximum available, 144
   maximum stable, 144
   nondiagonal SQUID, 237
Generator:
   airborne, 257
   aircraft, 251
   superconducting, 245
Gradiometer:
   first-order, 221
   portable, 235, 237
   second-order, 231, 234
   SQUID-based, 214
Grain:
   alignment, 14
   boundary, 9, 10, 11, 20
   orientation, 36
   texturing, 24

Heat:
   exchanger, 288
   impact, 283

load, 283
mass flow, 291
regenerator, 282
Heat exchanger:
counterflow, 289–290
regenerative, 289
Helium:
boil off, 277
cooling, 247
consumption, 252
liquid, 284
Homopolar motor:
rail transportation, 258
ship propulsion, 258
HTSC:
coil, 17
technology, 17
wire, 17
Hybrid design, 98

Imaging:
night vision, 81
sensors, 148
thermal, 81
Impact resistance, 125
Infrared (IR):
detector, 189
focal planar array, 207
forward-looking IR (FLIR), 189, 206
imaging camera, 189, 204
imaging sensors, 205
IR line scanner (IRLS), 206, 287
IR search and track (IRST), 287
military, 287
missiles, 288
quantum well (QW), 190
radiation, 196
receiver, 207
region, 192
staring camera, 205
surveillance, 275
tracking systems, 288
Integration:
DJI phase shifters, 95
MMIC technology, 38
Interaction, gyromagnetic, 98, 105

Junction:
capacitance, 195
Josephson, 12
noise, 190
temperature, 210

Kinetic:
energy, 26
inductance, 30

Laser:
crystals, 200
diodes, 189
double-quantum well (DQW), 199
multiple-quantum well (MQW), 199
pumping source, 200
radar, 200
ring, 190
semiconductor, 199
single-quantum well (SQW), 199
solid state, 189
superconducting, 199
Lattice:
vibration, 15
vortex, 14
Levitation:
system, 264
technique, 264
train, 264
Liquid nitrogen, 17
London penetration depth, 73, 79, 99, 166, 194
Loop:
current/voltage bias, 221
inductance, 215
impedance, 215
SQUID, 215
Losses:
conductor, 54
dielectric, 27
eddy-current, 249
gas leakage, 294
insertion, 26
magnetic, 15, 18
ohmic, 82
radiation, 82
reflector, 54
scattering, 29
substrate, 54
thermal, 277, 285
transmission line, 27

Machine:
airborne superconducting, 257
efficiency, 250
electrical, 242
mechanical, 242
round rotor, 242
synchronous, 242
Magnetic:
angle, 15
biasing field, 102
coil, 99, 285
coupling, 238, 249
field, 105, 212, 224
field resolution, 221

304 INDEX

Magnetic (*Continued*)
  flux, 14, 18, 90, 215
  flux exclusion, 89
  flux noise, 225
  ions, 56
  interaction, 55
  levitation, 262
  moment, 105, 212
  path, 99
  shielded room, 217
  vector, 232
Magnetic resonance:
  imaging (MRI), 21
  therapy (MRT), 238
Magnetic resonance imaging (MRI):
  LTSC, 17
  magnetic noise, 292
  medical application, 292
  patient comfort, 292
  sensor, 26, 214
  system, 278
Magnetometer:
  multichannel SQUID, 231
  SQUID, 214
Mapping:
  electrocardiogram (ECG), 233
  electroencephalogram (EEG), 233
  magnetocardiogram (MCG), 233
  neural activity, 215
Mean-time-between-failure (MTBF), 123
Mechanical:
  connectivity, 22
  integrity, 98
  properties, 24, 124
  strength, 111
Meissner effect, 37
Microcooler, 288
Microwave:
  cavity, 30, 57
  circuit, 55
  components, 187
  delay lines, 161
  devices, 26, 187
  field, 55
  filters, 159
  frequency, 98
  monolithic, 99
  power, 57
  range, 50
  switch, 114
Microstrip:
  structure, 70
  transmission line, 64

Microstructure, 18, 23
Millimeter (MM) wave:
  frequency, 47, 50, 53–54, 198
  gyrotrons, 111
  missile seekers, 148, 198
  telescopes, 207
  TWTA, 170
Mixer:
  Schottky diode, 207
  SIS quasi-particle, 207
MMIC technology, 38
Motors:
  air-cooled synchronous, 262
  conventional, 258
  dc homopolar, 257
  high torque, 274
  superconducting, 257, 260
Multichannel:
  SQUID, 231
  systems, 231

Neurological:
  aspects, 218
  disorders, 216
Neuromagnetic:
  applications, 238
  data, 231
  diagnosis, 230
  field, 235, 237
  laboratory, 231
  measurements, 235
  performance, 225
  rsearch, 231
  signals, 228
Noise:
  ambient, 233
  figure, 136
  Josephson, 209
  phase, 164
  quantum, 209
  shot, 205, 209
  temperature, 152
Normal:
  conductor, 34
  electron, 35
Nuclear magnetic resonance (NMR), 235

Ocean:
  current-monitoring, 17, 285
  floor-mapping, 17, 285
Operating:
  fequency, 34
  temperature, 31, 35

Optical:
  communication, 200
  components, 26, 189
  confinement, 201
  data link, 190
  delay line, 190
  detector, 20, 189
  fiber optic, 190
  modulation, 210
  power, 202
Oscillators:
  crystal, 166
  hybrid, 169
  microwave, 166
  phase noise, 128, 142
  SAW, 166
  YIG-tuned, 185

Penetration, London, 73, 79, 99, 166
Permeability tensor, 105
Phase:
  formation, 18
  noise, 122, 164, 166
  shift, 93
  transformation, 7
Phase shifters:
  DJI, 89
  ferroelectric, 95
  hybrid, 95
  switched-line, 114
  true time-delay, 89
Polarization:
  circular, 99
  linear, 82
  magnetic, 99
Power:
  consumption, 148
  handling capability, 26
  spectral density, 217
Propulsion systems:
  levitated, 241
  ship, 241

Quality factor:
  conductor, 49
  stripline, 26
  substrate, 27, 49, 117
  transmission line, 78, 102

Radar:
  FM-CW, 185
  mechanically-scanned, 186
  phased-array, 185–186
  signal processor, 186
  systems, 185

Radiation:
  efficiency, 153
  resistance, 153–154
  shield, 282
Radio frequency (rf):
  active, 88
  drive, 34
  SQUID, 215
  surface resistance, 22, 27, 34–35
Rare-earth material, 288
Receivers:
  acosto-optic, 174
  channelized, 173
  coherent, 207
  compressive, 174
  ESM, 174
  mixer, 207
  quasi-optical SIS, 209
  radar, 161
  sensitivity, 173
  surveillance, 177
Refrigerant, 245, 267, 298
Refrigeration:
  capacity, 283
  requirements, 266
  systems, 238, 245, 267
Refrigerators:
  closed-cycle, 277
  commercial, 276
  DM, 277
  GM, 277
  magnetic, 276
  Stirling, 277, 292
  two-stage, 277
Reliability:
  interconnection, 98
  long-term, 126
  MTBF, 211
  substrate, 46
Resistance:
  radiation, 154
  surface, 155
Resistivity, 54
Resolution:
  spatial, 231
  temporal, 231
Resonators:
  coplanar waveguide (CPW), 79
  microstrip, 79
  ring, 78, 166
  superconducting, 78
Responsivity, 81

Saturation velocity, 134

Scattering effects, 46
Sensors:
   astronomical, 198
   imaging, 148
   radiometric, 164
   space reconnissance, 95, 161
Shield:
   axial damper, 247
   damper, 245
   heat, 267
   magnetic, 247
   radiation, 247
   thermal radiation, 245
Shielding effectiveness, 247
Signal:
   brain, 230
   brain magnetic, 226
   evoked, 217
   eye, 230
   heart, 230
   modulation, 220
   neuromagnetic, 228
   processor, 205
   S/N ratio, 205
   stomach, 230
   vertical magnetic, 233
Signal processing devices:
   CCD, 79
   digital, 79
   IC, 79
   optical, 79
   SAW, 79
Solid state:
   devices, 116
   diode pump laser, 202
   microwave, 116
   MM-wave, 116
   varactor, 117
Sonar transmitter, 285, 287
Space:
   communication system, 184
   imaging sensor, 148
   sensors, 95, 166
   system, 142
Spatial:
   resolution, 212
   sensitivity, 212
Spectral:
   analysis, 174, 176
   density, 150
   purity, 164
   range, 190
   response, 80, 197
Spurious levels, 180, 182

SQUID:
   bandwidth, 217
   devices, 88, 215
   digital, 214, 225
   geometry, 90
   gradiometer, 218
   head, 220
   loop, 215
   LTSC, 178
   magnetometer, 177–178, 214
   microscope, 212
   RF-DC, 215
   sensitivity, 216, 218
   three-dimensional vector, 222
Substrate(s):
   anisotropic, 41
   ceramic, 46, 95
   dielectric, 28
   foam, 44
   granularity, 28
   hard, 28, 45
   metallic, 31, 38, 40, 57
   soft, 7, 41
   thin, 47
   undoped, 136
Subtrate properties:
   coefficient of linear expansion (CLE), 49
   dimensional stability, 42
   electrical, 40
   ferromagnetic, 100
   insertion loss, 51, 54
   loss tangent, 43
   mechanical, 40
   permittivity, 42
   physical, 40
   shear modulus, 48
   soft, 40
   tensile modulus, 46
Superconducting (SC):
   ADC, 177
   airborne generator, 258
   array, 156
   bearing, 269–270, 273
   cables, 266
   circuits, 46
   coherence length, 9
   components, 5, 46
   compounds, 3
   current transfer switch, 269, 271
   devices, 56
   DRFM, 182
   electrodes, 210
   electrodynamic levitation system, 262
   electronics, 173

fault current limiter (FCL), 269, 271
films, 9–10
filament matrix, 255
high-speed trains, 264
high-temperature SC (HTSC), 10
low-temperature SC (LTSC), 14
machines, 242
microstrip, 156
properties, 11
resonators, 78
shield, 78
ship-propulsion, 287
switch, 114
technology, 17
transmission lines, 266
Superconductor:
  ceramic, 241
  films, 20
  forms, 20
  metallic, 241
  ribbons, 20
  single crystal, 36
  skin-depth, 193
  tapes, 20
  wires, 20
Switching:
  packet, 183
  speed, 97
  time, 114
Synchronous:
  ac generator, 257
  ac motor, 257
  generator, 245
  machines, 242, 245
  ship propulsion, 258
Systems:
  cellular-based, 184
  communication, 80, 82, 88, 124, 184
  DRFM, 182
  electronic warfare (EW), 82, 88
  masers, 170
  missile seeker, 119
  MM-wave, 88
  nuclear fusion, 115
  optical, 88
  radar, 80, 82, 88, 95
  radiometric, 184
  ship-propulsion, 251
  space, 142

Technique:
  electronic countermeasure (ECM), 182
  superconducting, 17

Technology:
  collector, 257
  cryocooler, 285
  focal planar array, 189
  HTSC, 3, 5, 7, 10, 14, 161
  IR, 177
  LTSC, 2–3, 5, 177
  LTSC-JJ, 178
  magnetostrictive, 285
  medical, 225
  microelectronics, 178
  MMIC, 161
  printed-circuit, 161
  processing, 179
  slip ring, 257
Temperature:
  annealing, 24
  channel, 135
  characteristic, 202
  critical, 14
  cryogenic, 102
  curie, 101
  junction, 210
  noise, 143, 147, 170, 207
  peak, 271
  sintering, 7, 22
  spatial, 297
  stability, 294
  transition, 3
Thermal:
  annealing, 126
  conductance, 198
  conductivity, 119
  dissipation, 184
  effects, 126
  efficiency, 39
  excursion, 271
  insulation, 267, 269
  isolation, 145
  mass, 186
  noise, 150
  performance, 119
  resistance, 121
  shock, 124
  stresses, 123
Thermodynamic:
  aspects, 289
  comparision, 289
  efficiency, 290
  stability, 34
Thermomechanical, 38
Threshold:
  current, 202
  density, 202

Time-of-arrival (TOA), 176
Transconductance (G_m):
  extrinsic, 136
  instrinsic, 130, 134
Transistors:
  DB-HBTs, 126
  flux flow, 126
  HBTs, 116, 125, 143
  HEMTs, 116
  inverted-HEMTs, 136
  MESFETs, 116, 128, 143
  MODFETs, 133
  p-HEMTs, 116
  pseudo-HOBs, 126
  three-terminal, 126
  transconductance, 128
Transmission lines:
  cooplanar waveguide (CPW), 114
  exponential, 76
  meander, 76
  microstrip, 76, 91
  overhead, 266
  power, 241
  printed circuit, 114
  superconducting power, 266
  variable velocity, 95
Transmitter power, 185
Tuning:
  coil, 102
  fast, 98
  fine, 98
  speed, 102
Turbogenerators:
  conventional, 241, 254
  high power, 274
  superconducting, 254

Two-fluid model, 64

Ultra-low noise, 172

Valve:
  needle, 297
  opening, 297
Vertical polarization, 187

Washer-type DC-SQUID, 228
Wavelength:
  longer, 201
  short, 202
  wide coverage, 201
Windings:
  armature, 242
  conventional, 242
  damper, 245
  field, 242
  harmonics, 251
  rotor, 242
  stator, 247, 249

X-Band:
  frequencies, 28
  GaAs MESFET amplifier, 143
  ring resonator oscillator, 166

YBCO:
  ceramic, 21
  compound, 4
  film, 11
  material, 6
  wire, 21
Yttria-stabilized-zirconia (YSZ), 28

Zirconia, 27, 53